Collection of Sand

ITALO CALVINO

Collection of Sand

Essays

Translated by MARTIN MCLAUGHLIN

Mariner Books
Houghton Mifflin Harcourt
BOSTON · NEW YORK

First U.S. edition

Copyright © 2002 by The Estate of Italo Calvino

English translation and additional editorial material
copyright © Martin McLaughlin, 2013

www.hmhco.com

First published in Italy as *Collezione di sabbia* by Garzanti, 1984

This translation first published in the United Kingdom by Penguin Classics, 2013

Library of Congress Cataloging-in-Publication Data
Calvino, Italo.
[Collezione di sabbia. English]
Collection of sand / Italo Calvino; translated by Martin McLaughlin. —
First U.S. Edition.
pages cm
ISBN 978-0-544-14646-4 (pbk.)
1. Calvino, Italo — Translations into Spanish. 2. Calvino, Italo —
Translations into English. I. McLaughlin, M. L. (Martin L.) translator.
II. Title.
PQ4809.A45C5513 2014
854'.914 — dc23   2014001373

Printed in the United States of America
DOC 10 9 8 7 6 5 4 3 2 1

Contents

Contents

Translator's Introduction

Collection of Sand was published in Italian as *Collezione di sabbia* in October 1984. It was the last organic volume of new work put together by Italo Calvino in his lifetime (the only book to appear after it and before the author's death in 1985 was the final anthology of cosmicomic stories which largely reproduced previously published tales and included only two new pieces). *Collection of Sand* was the second substantial collection of non-fiction that the author had published. The previous book of essays entitled *Una pietra sopra* (*Moving On*) had appeared in 1980 and had contained writings on politics, society and literature. Calvino chose that first title – *Una pietra sopra* means literally 'putting a stone over something', in other words drawing a line under something – because by 1980 he considered that a certain phase of his life, that of the committed left-wing intellectual, had by now come to an end and that the world was too complex to be changed by the kind of attitudes that had sustained him in his youth and maturity. His early aspirations to write literature that would somehow lead to a new society foundered as it became clear that the intellectual had very little power to influence events in the 1960s and 1970s (the latter decade in Italy had been characterized by unparalleled levels of political terrorism, culminating in the Red Brigades' capture and killing of the Christian Democrat prime minister Aldo Moro in 1978 and the neo-Fascist bombing of Bologna railway station in 1980).

Critics have noted how Calvino's fascination with mineral imagery is reflected in the Italian titles of these two substantial essay collections (stone, sand), and there is no doubt that the image of rocks eventually turning to sand, emblematic of the slow passing of geological eras, is a cosmic theme apparent in the writer's other works. However, the differences between the two volumes are even more significant. The unitary stone of the first title has given way to the granules of sand in the second, the latter image reflecting the author's own fragmentation and perplexity in the face of an ever more rapidly changing world. In *Collection of Sand* the committed intellectual of the first volume of essays, who had written on society and literature, has disappeared to be replaced by someone who is quite removed from the Italian socio-political-literary scene: he is now a reviewer of exhibitions, an observer of art-works and art-books, and a traveller in non-European countries (Japan, Mexico, Iran). Calvino here claims to write as a dilettante (he is not an art historian nor an experienced travel-writer), yet, as he says in the blurb to the first edition of the collection, what is on display in these essays is his encyclopedic curiosity and his appetite for meticulous observation in trying to understand what he calls 'the truth of the world'. One implicit message here is that the discreet observer of art and of other countries can perhaps offer as much as the committed intellectual to the reader trying to understand societies and cultures.

The volume is divided into four main parts. First, there is a section entitled 'Exhibitions – Explorations', containing ten reviews, mostly of exhibitions Calvino had seen in Paris in 1980–84. The second part, 'The Eye's Ray', consists of eight pieces from the same period, devoted to aspects of the visual, from an essay on Roland Barthes's book on photography, and a virtuoso ekphrasis of Trajan's Column, to a review of a book on the history of optics and of ideas on how the eye sees. Section III, 'Accounts of the Fantastic', also from the early 1980s, moves to the imaginary world, and comprises five book reviews, ranging from a seventeenth-century Scottish treatise on the geography of fairies to works by remarkable contemporary artists of visual fantasy such as Donald Evans and Luigi Serafini. The fourth and final section, 'The Shape of Time', consists of fifteen travelogues, nine devoted to Japan and three each to Mexico and Iran, the Japanese and Mexican pieces being written in 1976 and the Iranian ones in 1975. As the final section's title suggests, these are not

just accounts of journeys crossing geographical space but also specu-
lations on larger questions of time and history. What unites each of
the thirty-eight essays thematically is Calvino's fascination with all
aspects of the visual universe and of seeing: not just what we see but
how we see and how we then interpret what we have seen. In terms
of genre, most of these descriptions of art-works and sites are examples of ekphrasis, verbal portraits of works of art, but it soon
becomes clear that at the same time as being a volume of ekphrastic
essays, this book is also the work of Calvino the narrator: many of
the reviews of exhibitions or art-works shift into engaging narratives (as in the pieces on Trajan's Column, on Delacroix's *Liberty
Leading the People*, and so on). If an ekphrasis was in origin a description of a work of art designed to provide a contrastive digression
from a main story (the first one in Western literature was Homer's
description of the shield of Achilles in the *Iliad*), Calvino offers us
here a whole series of ekphrases, but all of them spill over into narrative, and many of them highlight his capacities as a writer of tales.
These are visual stories as well as essays on sight.

The components of *Collection of Sand* might be regarded as examples of 'late Calvino' (he was to die in September 1985, under a year
from the publication of this book), but they are all concerned with
quintessentially Calvinian themes that figure in different parts of his
oeuvre. His fascination with the visual is apparent in all his works:
talking of the genesis of the trilogy *Our Ancestors* he claimed that all
his narratives started out from a visual image in his head. In these
essays the image is offered to his imagination by an exhibition, a
book or a place, but they lead to elegant, thoughtful narratives
here too: as he says in one essay, 'the brain begins in the eye'. They
are particularly reminiscent of the short pieces that make up *Mr
Palomar*, Calvino's fictional work of 1983 (and a number of them
were written at exactly the same time), each of which – in the words
of Seamus Heaney – feels like a single inspiration being caught just
as it rises and being played to explore its maximum potential. Like
Mr Palomar, here too the author manifests his concerns with waste,
entropy and looming catastrophe. Unsurprisingly, many of the
essays are also about writing, about when script first emerged, about

using other semiotic systems such as knots, about how writers turn to drawings in dissatisfaction with the written word, ideas explored throughout Calvino's oeuvre.

These are essays that embrace both the world and the word. But as well as the thematics of stone and sand implicit in the title and evident in many essays, the author's preoccupation with the world of nature, and in particular trees, stands out. Calvino's agronomist father Mario had lived in Mexico and Cuba in the early years of the century, advising the inhabitants on agriculture and plant-growing: indeed he was in Cuba when the young Italo was born there in 1923. Trees and plants figure regularly in Calvino's fiction: the author's great hymn to the arboreal world is to be found in *The Baron in the Trees*, especially chapter 10, with its precise description of the leaves and bark of the different kinds of trees that made up the hero's habitat. But here in the genre of non-fiction we find trees everywhere, often accompanied by similarly meticulous descriptions: they appear in the essay on the New World, in the section in 'The Traveller in the Map' on the precise number of trees in French forests, in the attempt at classifying the trees represented on Trajan's Column, in the loving descriptions of the maple and ginkgo trees in Japan, and at the end of the volume in the detailed description of the enormous 'Tule Tree' in Mexico. This luxuriant arboreal theme stands in clear contrast to the mineral imagery of stone and dust.

Collection of Sand brings together many motifs that resonate elsewhere in Calvino's work. The detail of the weapons that have changed sides in Delacroix's *Liberty Leading the People* is one that fascinated the writer, since he had experienced it at first hand in his partisan days, had read about it in his beloved Ariosto, and wrote about it in his first novel. His enthusiasm for the Enlightenment, which had surfaced in narrative form in *The Baron in the Trees*, is present here too in the many mentions of the eighteenth century and its achievements. Similarly the moon, one of his favourite images, surfaces here in several pieces, often with echoes of Leopardi (especially in 'The Moon Chasing the Moon'). But Calvino is not just fascinated by his own times and the recent past. His interests in classical art and archaeology in the essays on Trajan's Column and on the dig at Settefinestre chime with the author's return to the Graeco-Roman classics in other seminal essays of the late 1970s and early 80s, notably those contained in *Why Read the Classics?* However, the fictional works that are most often evoked by these essays are *Invisible Cities* and *Mr Palomar*. There are several mentions of Venice, discussions of ideal and imaginary cities and paradoxical statements reminiscent of the rewriting of Marco Polo's travels: as when the descriptions of the bare and unadorned imperial palaces in Japan make the author presume that these can exist only because there are other houses in the country 'chock-full of people and tools and junk and rubbish, with the smell of frying, sweat, sleep, houses full of bad moods, people rushing, places where people shelled peas, sliced fish, darned socks, washed sheets, emptied bed-pans' ('The Obverse of the Sublime', 1976). In fact, the visit to Japan seems to have been an extremely fertile trip in creative terms as one particular moment inspired two totally different kinds of writing: the description of the ginkgo tree's leaves falling on the ground is present in one of the travelogues here ('The Obverse of the Sublime'), but a few year's later in *If on a Winter's Night a Traveller* the same scenario of falling ginkgo leaves resurfaces as the opening of the sensual Japanese micro-novel in that 1979 work (also inspired by the brief essay here on Japanese erotic prints). Observation of a detail of nature can spark the author's imagination into either fictional or non-fictional creativity.

The three countries that form the subject of the final section of this book – Japan, Mexico, Iran – are the three countries visited by the eponymous 'hero' in a central section of *Mr Palomar*: once more, the reader can compare the description of the sand-garden of the Ryoanji temple here with its counterpart in that fictional work and contrast the two different ways in which the narratives develop. Similarly, the description of the Aztec temple being overgrown by the forest in this volume of essays is echoed in narrative terms both in the Mexico story in *Mr Palomar* and in the title story of *Under the Jaguar Sun*. Calvino's fascination with the cultures of these three non-Western countries is partly explained by the fact that he was haunted by the notion that literature and culture must avoid limits, hence his interest during the 1970s and 80s both in systems of thought that were not those of the European West, and in every aspect of the visual universe. In short, these are essays on what is termed in *Mr Palomar* the inexhaustible surface of things.

One of the great pleasures of this collection is enjoying Calvino's inimitable style as an essayist. He was always a passionate advocate of stylistic *brevitas*, and he loves to begin essays with a brief, limpid sentence. There are some outstanding examples here of short, one-clause opening sentences: 'There is a person who collects sand'; 'Their first attribute is lightness'; 'In the beginning was language.' However, on other occasions, by way of contrast, he opens with a lengthy sentence, which is never gratuitous but usually reflects the subject-matter, as for instance the first sentence describing the Mihrab, or the one introducing the Mexican temple smothered by the jungle: 'In Palenque the soaring temples built on steps stand out from the background of the forest that rises above them with dense trees that are even higher than the temples: *ficus* trees with multiple trunks that look like roots, *aguacetes* with their shiny leaves, cascades of creepers, dangling plants and lianas' ('The Forest and the Gods').

Calvino is also a master of the closing sentence too: good examples are the final words of the essay on Delacroix ('After a number of other vicissitudes, under the Third Republic, the work entered the Louvre, and after that into universal glory'), or the one on archaeology ('The archaeologist's spade and trowel try to reconstruct the continuity of history through the long intervals of darkness'). Elsewhere than in the opening and closing sentences we find plenty of instances of his mastery of prose style, such as his love of paradoxes (see the passage on the 'crowded solitude' of Japanese *pachinko* or pinball arcades) and his fondness for images of lightness, for instance his summing up of Donald Evans's stamps project, which is described as 'a ritual of private celebrations, commemorations of minimal encounters, consecrations of things that are unique and irreplaceable: basil, a butterfly, an olive'. These striking images conjure up the author's cult of lightness, as expressed in the first Harvard lecture on the subject, in the posthumous *Six Memos for the Next Millennium*.

The essays in *Collection of Sand* may seem occasional and hetero-geneous but they all share thematic concerns around sight and writ-ing, and there are many links between individual essays. The motif of sand which appears in the first piece resurfaces in the dust into which Trajan's Column is disintegrating, reappears in one of the essays towards the end about the lunar sand in a Japanese Zen gar-den, and makes a last appearance in the final essay on the stone sculptures in Persepolis and Naqsh-e Rustam in Iran. Similarly the Japanese and Aztec civilizations which dominate the final section of the volume appear as early as the third essay on 'The Traveller in the Map'. One final overarching motif is that of photography, which crops up in several pieces, from the photographs of crimes in the essay on the popular press, to the reflections on photography in the article on Barthes to the description of camera-happy Japanese tourists in Mexico, in one of the final pieces in the volume.

Calvino was a prolific essayist and reviewer. His output in non-fiction matches his voluminous output in fiction, but so far in English only four collections have appeared: *The Literature Machine*, which appeared in Calvino's lifetime (1982), contained a small num-ber of essays from *Una pietra sopra* plus other essays from a range of different sources; the other non-fiction works in English were the three posthumous collections of essays, *Six Memos for the Next Millennium* (1988), *Why Read the Classics?* (1999), and the autobio-graphical *Hermit in Paris* (2003). *Collection of Sand* is the first transla-tion in English of a substantial volume of essays put together in a specific order by the author himself during his own lifetime. It is a book of variety but also unity, as we have seen, and it resembles a verbal tour of exhibitions, art-works and countries in the company of a stimulating commentator.

Collection of Sand is, appropriately enough, a granular book, the individual pieces being like grains of sand in a collection. In fact Calvino chose the first essay as the title essay of the whole volume because it was for him a manifesto piece. Like the opening story in *Mr Palomar* ('Reading a Wave'), it was deliberately placed first in the book since its ideas range way beyond the object observed in an attempt to encompass the universe. Indeed the occasion for the writing of the title essay here was a slightly vertiginous exhibition, a collection of collections, but the one that fascinated the author in that exhibition was the collection of sand. In this potentially eccentric hobby Calvino sees something much more profound and significant, something that records what remains of the world after centuries-long erosions, something that is 'both the ultimate substance of the world and the negation of its luxuriant and multiform appearance'. However, in addition to this cosmic dimension, he sees in the collection of sand something more personal, a metaphor for the author's oeuvre, as he asks himself 'what is expressed in that sand of written words which I have strung together throughout my life, that sand that now seems to me to be so far away from the beaches and deserts of living'. The final essay on the stone sculptures of the Achaemenid and Sassanid kings in Iran is deliberately placed in a position of closure, as it speculates on how to escape time and develops a thoughtful contrast between the permanence of those imperial reliefs and the seasonal wanderings of the nomads who inhabit those regions. It is no accident that one of the words that recurs most frequently in these essays, dust, reappears in this final essay to mark a contrast between the rulers immortalized in stone and the nomads: 'For centuries the nomads have criss-crossed these arid terrains between the Persian Gulf and the Caspian Sea without leaving any trace of themselves behind apart from their footprints in the dust.' The problem for powerful rulers, whether they are Trajan or Darius or Shapur I, is that their proud monuments will all turn to dust. The writer's sense of how to construct and end a collection is apparent here, just as it is in his fictional works. Reading *Collection of Sand*, the English reader will discover that Calvino the essayist is every bit as intriguing and satisfying as Calvino the novelist.

Martin McLaughlin, 2013

The translator would like to thank the following people for their expert help in solving a number of particular problems: Catherine McLaughlin, Graham Nelson, Michael Sheringham, Giuseppe Stellardi, Elisabetta Tarantino, David Watson.

The Author's Presentation of the Volume

From Paris Italo Calvino periodically sends an article on an unusual exhibition to the newspaper he collaborates with [la Repubblica].* This allows him to tell a story through a series of objects: ancient maps or globes, wax manikins, clay tablets with cuneiform writing, the popular press, traces of tribal cultures and so on. Some traits of the author's physiognomy come through in these 'occasional' pieces: an omnivorous, encyclopedic curiosity, and a desire to distance himself discreetly from any form of specialism; respect for journalism as a way of providing impersonal information and the pleasure of relegating his own opinions to marginal observations or hiding them between the lines; an obsessive meticulousness and dispassionate contemplation when it comes to the truth of the world. Along with ten of these accounts of walks through the rooms of Parisian galleries, Collection of Sand includes other essays on 'things observed', or essays which, even though they start out from the reading of a book, have as their subject the visible or the very act of seeing (including what the imagination sees). The volume is rounded off by three sections containing reflections penned during his travels to other civilizations: Iran, Mexico, Japan. These pieces start from 'things observed' and open out to offer glimpses of other dimensions of the mind.

* This anonymous note was written by Calvino for the back cover of the first edition of Collezione di sabbia (Milan: Garzanti, 1984).

A Note on the Text

The essays in parts I, II and III were all published in *la Repubblica* between 1980 and 1984, apart from the following: 'Collection of Sand' (*Corriere della Sera*, 25 June 1974); 'How New the New World Was' (a spoken commentary for broadcast by RAI-TV, December 1976); 'The Encyclopedia of a Visionary' (*FMR*, vol. I, March 1982).

The fourth part, 'The Shape of Time', includes pieces on Japan and Mexico, from 1976, partly published in *Corriere della Sera* and partly unpublished, and passages on Iran that have never been published before, from notes made during a journey there in 1975.

Collection of Sand

I Exhibitions – Explorations

Collection of Sand

There is a person who collects sand. This person travels the world and – on arrival at a sea-shore, the banks of a river or lake, or a desert, or wasteland – gathers a handful of sand and takes it away. On returning home, thousands of little jars are waiting, lined up on long shelves: inside them the fine grey sand of Lake Balaton, the brilliant white particles from the Gulf of Siam, the red shingle that the Gambia river deposits on its course down through Senegal, all display their not particularly vast array of nuanced colours, revealing a uniformity like the moon's surface, despite the differences in granulosity and consistency, from the black and white sand of the Caspian Sea, which seems to be still bathed in salt water, to the tiniest pebbles from Maratea, which are also black and white, to the fine white powder speckled with purple shells from Turtle Bay, near Malindi in Kenya.

In a recent Paris exhibition about bizarre collections – collections of cowbells, bingo games, bottle-tops, terracotta whistles, train-tickets, spinning-tops, toilet-paper packaging, collaborators' badges during the German occupation, embalmed frogs – the case with the collection of sand was the least showy but also the most mysterious, the one that seemed to have most things to say, even through the opaque silence imprisoned behind the glass of the jars.

Surveying this anthology of sands, the eye initially takes in only the samples that stand out most, the rust-coloured sand from a dry river-bed in Morocco, the carboniferous black and white grains from the Aran Islands, or the shifting kaleidoscope of reds, whites, blacks and greys that has on its label a name that is even more polychromatic:

Parrot Island, Mexico. After this, the minimal differences between one kind of sand and another demand a level of attention that becomes more and more absorbing, so much so that one enters into another dimension, into a world that has no other horizons except these miniature dunes, where one beach of tiny pink pebbles is never the same as another beach of tiny pink pebbles (they are mixed with white grains in Sardinia and in the Grenadine Islands in the Caribbean, whereas they blend with grey grains in Solenzara on Corsica), and a sample of minuscule black shingle from Port Antonio in Jamaica is not the same as one from Lanzarote in the Canary Islands, nor as another that comes from Algeria, maybe from the middle of the desert.

One has the feeling that this set of samples from the universal Waste Land is on the point of revealing something important to us: a description of the world? The collector's secret diary? An oracular response to myself as I scrutinize these motionless sand-clocks and reflect on the moment I have reached in my life? Maybe all of these things together. The collection of sands that have been selected chronicles what remains of the world from the long erosions that have taken place, and that sandy residue is both the ultimate substance of the world and the negation of its luxuriant and multiform appearance. All the scenarios of the collector's life appear more alive here than if they were in a series of colour slides. In fact, one would think that this sounds like a life of eternal tourism (and that is just the way that life appears in any case in colour slides, and it is how posterity would reconstruct it if it were only slides that remained to document our times): basking on exotic beaches is alternated with more arduous explorations, in a geographic restlessness that betrays a sense of uncertainty, of anxiety. Such scenarios are evoked and at the same time cancelled out by what is the by now compulsive action of bending down to collect a little bit of sand and filling a little bag with it (or a plastic container? Or a Coca-Cola bottle?) and then turning round and leaving.

Actually, just like every collection, this one is a diary as well: a diary of travels, of course, but also of feelings, states of mind, moods, even though we cannot be sure that there really is a correlation between on the one hand the cold, earth-coloured sand from Leningrad, or the very fine sand-coloured shingle from Copacabana, and on the other the feelings the sands arouse when we see them bottled and labelled here. Or perhaps it is only a record of that obscure mania which urges us as much to put together a collection as to keep a diary, in other words the need to transform the flow of one's own existence into a series of objects saved from dispersal, or into a series of written lines abstracted and crystallized from the continuous flux of thought.

The fascination of a collection lies just as much in what it reveals as in what it conceals of the secret urge that led to its creation. Amongst the weird collections in the exhibition, one of the most alarming was clearly the collection of gas-masks: out of its case stare green or greyish faces made of canvas or rubber, with round, blind staring eyes, and the snout-like nose like a tin or a supple tube. What spirit motivated this particular collector? A sense of irony but also, I believe, a feeling of terror regarding a humanity willing to adopt facial features somewhere between the animal and the mechanical. Or perhaps also a confidence in the resources of anthropomorphism, which invents new forms in the image and likeness of the human face, in order to equip itself to breathe in phosgene or mustard gas, and at the same time not without a hint of fun and caricature. And certainly also a sense of revenge against war, by showing in these masks the rapid obsolescence of this aspect of war, and thus rendering it more ridiculous than terrifying. But also the sense that amidst war's shocking, foolish cruelty one can still make out our own true image.

Certainly, if the collection of gas-masks managed somehow to transmit a mood that was in some way amusing and heartening, just a little further on a chilling, anguished effect was produced by a collector of Mickey Mouse products. Someone has collected, clearly throughout a whole lifetime, dolls, boxes of toys, berets, masks, t-shirts, furniture, bibs, all of which reproduce the stereotypical features of Disney's Mickey Mouse. From this packed display-case hundreds of round black ears, of white faces with a black spot for a nose, of long white gloves and skinny black arms reveal a childish obsession with that single reassuring image in the midst of a terrifying world, so that in the end this one mascot, in its countless mass-produced manifestations, itself becomes imbued with a feeling of terror.

But where the collecting mania turns in on itself, revealing its own underbelly of self-obsession, is in a case full of plain cardboard covers, tied with ribbons, on each of which a female hand has written titles such as *Men I Like*; *Men I Don't Like*; *Women I Admire*; *My Jealousies*; *My Daily Shopping*; *My Fashion Tastes*; *My Childhood Drawings*; *My Castles*; and even: *Paper-Wrappers from the Oranges I Have Eaten*.

What these folders contain is not a mystery, because this is not someone who exhibits occasionally but a professional artist (Annette Messager, Collector: that is how she signs herself), who has staged various solo exhibitions in Paris and Milan, based on her series of newspaper cuttings, pages of notes and sketches. But what is interesting in this collection is precisely this expanse of closed, labelled covers and the mental procedure they imply. The author herself has clearly defined it: 'I try to possess and appropriate life and the events I get to know of. Throughout the whole day I leaf through things, collecting, ordering, classifying, sifting, and reducing everything to the form of so many collectors' albums. These collections then become my own life, illustrated.'

Her own days, minute by minute, thought by thought, reduced to a collection: life ground down to a dust-cloud of tiny grains: sand, once more.

I retrace my steps, towards the case with the collection of sand. The real secret diary to decipher is here, amidst these samples from beaches and deserts now under glass. In this case too the collector is a woman (as I read in the exhibition catalogue). But just now I am not interested in giving her a face, or features; I see her as an abstract person, an 'I' that could be myself as well, a mental mechanism which I try to imagine at work.

Here she is back from a journey, adding new containers to those already lined up, and suddenly she notices that without the indigo of the sea the sparkle of that beach of shattered shells has been lost; that none of the damp heat of the wadi has remained in the blobs of sand; that far from Mexico, the sand mixed with lava from the volcano Parícutin is just black powder that looks as if it has been swept down from a chimney. She tries to recall the sensations of that beach, that forest smell, that heat, but it is just like shaking that little bit of sand at the bottom of the labelled jar.

At this point there is nothing left to do except to give up, walk away from the case, from this cemetery of landscapes reduced to a desert, this cemetery of deserts on which the wind no longer blows. And yet, the person who has had the persistence to continue this collection for years knew what she was doing, knew where she was trying to get to: perhaps this was her precise aim, to remove from herself the distorting, aggressive sensations, the confused wind of being, and to have at last for herself the sandy substance of all things, to touch the flinty structure of existence. That is why she does not take her eyes off those sands, her gaze penetrates one of the phials, she burrows into it, identifies with it, extracts the myriads of pieces of information that are packed into a little pile of sand. Each bit of grey, once it has been deconstructed into its light and dark, shiny and opaque, spherical, polyhedral and flat granules, is no longer seen as a grey or only at that point begins to let you understand the meaning of grey.

So, deciphering the diary of the melancholic (or happy?) collector, I have finally come round to asking myself what is expressed in that sand of written words which I have strung together throughout my life, that sand that now seems to me to be so far away from the beaches and deserts of living. Perhaps by staring at the sand as sand, words as words, we can come close to understanding how and to what extent the world that has been ground down and eroded can still find in sand a foundation and model.

[1974]

How New the New World Was

Discovering the New World was a very difficult enterprise, as we have all been taught. But even more difficult, once the New World was discovered, was *seeing* it, understanding that it was *new*, entirely *new*, different from anything one had expected to find as *new*. And the question that spontaneously arises is: if a New World were discovered now, would we be able to *see it*? Would we know how to rid our minds of all the images we have become accustomed to associate with the expectation of a world different from our own (images from science fiction, for instance) in order to grasp the real difference that would be presented to our gaze?

We can instantly reply that something has changed since the time of Columbus: in the last few centuries man has developed a capacity for objective observation, a scrupulousness about precision in establishing analogies and differences, a curiosity for everything that is unusual and unexpected, and these are all qualities that our predecessors in the ancient world and in the Middle Ages apparently did not possess. It is precisely from the discovery of America, we can say, that the relationship with what is new changes in human consciousness. And it is for that very reason that we usually say that the modern era began then.

But will it really be like this? Just as the first explorers of America did not know at what point they would either be proved wrong or have their familiar preconceptions confirmed, so we too could walk past things never seen before without realizing it, for our eyes and minds are used to selecting and cataloguing only that which responds to tried and tested classifications. Perhaps a New World opens up every day and we don't see it.

These thoughts came to mind while visiting the exhibition America Seen by Europe, an exhibition that brings together more than 350 paintings, prints and objects at the Grand Palais in Paris. All of them represent European images of the New World, from the earliest reports that came back after the voyage of Columbus's caravels to the gradual understanding that emerged from accounts of the exploration of the continent.

These are the shores of Spain: it was from here that King Ferdinand of Castille gave orders for the caravels to set sail. And this stretch of sea is the Atlantic Ocean which Christopher Columbus crossed to reach the fabled islands of the Indies. Columbus leans out from the prow and what does he see? A procession of naked men and women coming out of their huts. Barely a year had passed since Columbus's first voyage, and this was how a Florentine engraver represented the discovery of what at that time people did not know would become America. Nobody yet suspected that a new era in the history of the world had opened up, but the excitement aroused by this event had spread throughout Europe. On his return, Columbus's report instantly inspired an epic in octaves in the style of a chivalric poem by the Florentine Giuliano Dati, and this engraving is in fact an illustration from that book.

The characteristic of the inhabitants of the new lands that most struck Columbus and all the early explorers was their nakedness, and this was the first detail that worked on the illustrators' imaginations. Men are portrayed as still having beards: the news that the Indians had smooth cheeks apparently had not yet spread. With Columbus's second voyage and especially with the more detailed and colourful reports by Amerigo Vespucci, another feature as well as their nakedness fired the European imagination: cannibalism.

Seeing a group of Indian women on the shore – Vespucci tells us – the Portuguese sent ashore one of their sailors, who was famous for his handsomeness, to talk with the Indian women. They surrounded him, lavishing embraces and expressions of admiration on him, but meanwhile one of their number hid behind his back and clubbed him on the head, stunning him. The unfortunate man was dragged away, cut into pieces, roasted and devoured.

The first question Europeans asked about the inhabitants of the New World was: are they really human? Classical and medieval traditions spoke of remote areas inhabited by monsters. But the lie was soon given to such legends: Indians were not only human beings, but specimens of classical beauty. That was how the myth arose of their happy life, unburdened by property or labour, as in the Golden Age or Earthly Paradise.

After the crude engravings on wood we find the depiction of Indians in paintings. The first American we see portrayed in the history of European painting is one of the Three Magi, in a Portuguese painting of 1505, in other words barely a dozen years after Columbus's first voyage, and even less time after the Portuguese landing in Brazil. It was still believed that the new lands were part of the Far East of Asia. It was traditional for the Three Magi in paintings of the Nativity to be represented in oriental garments and head-gear. But now that the explorers' reports provided direct evidence of how these legendary inhabitants of 'India' looked, painters brought themselves up to date. The Indian Magus was portrayed as wearing a feather headdress, as certain Brazilian tribes do, and carrying in one of his hands a Tupinambá arrow. Since this was a painting for a church, this character could not be portrayed naked: he has been given a Western waistcoat and trousers.

In 1537 Pope Paul III declared: 'The Indians are truly human . . . not only are they able to understand the Catholic faith, but they are extremely keen to receive it.'

Feather headdresses, weapons, fruits and animals from the New World started arriving in Europe. In 1517 a German engraver drawing a procession of inhabitants from Calcutta, mixes Asiatic elements (such as an elephant and its mahout, bulls draped in garlands, rams with huge tails) with details that come from the recent discoveries: a feather headdress (and actually clothes made of feathers that are totally imaginary), an Ara parrot from Brazil and also two corn cobs – maize was the cereal that was destined to play such a major role in the agriculture and diet of Northern Italy, but its American origin would be soon forgotten since the Italian word is 'granturco' [literally Turkish grain].

It is thanks to the work of the great cartographers of the sixteenth century that we see not only the new territories taking shape, but also the fauna, flora and customs of these peoples giving us their first true images. Working at close quarters with the explorers, the map-makers had access to information at first hand. The outlines of the Atlantic coasts were largely known when the new lands were still thought to be an appendix of Asia. Thus in a silver globe of 1530 the Gulf of Mexico is called 'The Sea of Cathay', and South America is 'Cannibal Land'.

It is in a German map that the name *America* appears for the first time, meaning 'Amerigo's Land', because it was mainly through the reports of Vespucci's voyages that Europe had taken on board the geographical significance of the discoveries. It was only after the arrival of the Florentine merchant's letters that Europe realized that what was opening up for the old continent was indeed a New World, of enormous dimensions and with its own characteristics.

Now suddenly in maps of the time America is detached from Asia. All that is known of North America (here called 'Land of Cuba') is a small strip of coastline, and it is thought to be near Japan (called 'Zipangri'). The name 'America' is applied only to Southern America, also called 'Terranova' and inhabited, of course, by cannibals. The continent has by now acquired an autonomous outline, but it is still seen – even in its shape – mainly as an obstacle, a barrier separating us from China and India.

In maps drawn up by Mercator, the inventor of a new method of cartographic projection, the name 'America' is now applied to the northern part of the continent as well, and the word is placed alongside Labrador, which was then called the Land of Cod.

The ideas people had of the Indians were polarized for a long time between two opposite myths: the myth of the natural happiness of an innocent life, as in the Garden of Eden, and that of ruthless ferocity: stories of flaying and torturing. But there were signs of a growing outrage at the cruelty of the Spanish, the slaughter and pillage carried out by the Conquistadors.

It is only towards the close of the sixteenth century that we really start to see Indians face to face. And this too was thanks to a cartographer and draughtsman, the Englishman John White, who in 1585 followed the expedition led by Sir Walter Raleigh, founder of the first English colony beyond the Atlantic, Virginia. White's seventy-six watercolours, now in the British Museum, constitute the first evidence of America drawn from life by a painter. White did not just draw the costumes and activities of the Red Indians, but also the animals of North America: the flamingos, iguanas, land-crabs, turtles, flying fish and the huge variety of aquatic fauna.

That America had a fauna and flora completely different from those of the Old World was a fact that took a long time to be recognized by Europeans. Right from his first voyage Columbus had brought back to Spain some parrots, Ara parrots, which were much bigger than African parrots. These aroused instant curiosity and were inserted by Raphael in the grotesque-style decorations in the Vatican Loggia.

But on the whole the new animals from America do not seem to have aroused much excitement. People soon began to rear turkeys in Europe, but they believed wrongly that they were of Asian origin, confusing them with guinea-fowl.

The animal that most caught the imagination was the armadillo, so much so that in allegorical representations America was portrayed as a naked woman, armed with a bow and arrows, riding on an armadillo.

The truth is that in this immense and fertile continent Europeans perhaps expected to find fauna of mastodontic proportions and were rather disappointed. America has plenty of strange animals but most of them are of modest size. That is why the makers of the Gobelin tapestries felt the need to add to their luxuriant vision of the flora and fauna of Brazil animals that have nothing to do with America. They contain the most typical zoological representatives of the New World, such as the anteater, the tapir, the toucan and the boa constrictor, but also an African elephant, an Asiatic peacock and a horse of the kind that the Europeans imported into America.

Just as slow, but with much more important consequences, was the conquest of Europe by American plants. The potato, the tomato, corn and cocoa, which were to have a key role in the agriculture and diet of the whole of the West, as well as cotton and rubber, which would dominate so much of our industrial production, and tobacco, which was to play such an important part in behavioural habits, all took a long time to be recognized as new plants. In the sixteenth century the study of nature was still based on Greek and Latin authors: it was not the new and the unusual that attracted scholars but only that which, rightly or wrongly, could be classified using the names handed down by classical texts.

In the exhibition we see a Flemish or German watercolour dated 1588 which has extraordinary historical value since it is the first known representation of a potato (which had been imported from Peru to Spain a few years previously), and a print which was the first illustration of a tobacco plant ever to be published, in 1574, in Antwerp. The small head of an Indian exhaling clouds of smoke through a strange, vertical pipe records that curious custom which no explorer had ever failed to note, and to which were attributed sometimes therapeutic and sometimes toxic properties.

In the seventeenth century it was the Dutch who, after hounding the Spanish out of Brazil and before being chased out in their turn by the Portuguese, sent scientists and artists to study nature in the colony. Albert Eckhout signals the meeting between Dutch nature and Brazilian vegetation. Water-melons, cashews, a custard-apple, a passion flower and a pineapple stand out against the sky like a mountain of tastes and perfumes. American pumpkins and cucumbers mingle with European cabbages and turnips in celebration of the unification of the world of vegetables on both sides of the Atlantic.

A painting by Franz Jansz Post, today in the Louvre, marks the moment when Dutch landscape painting comes into contact with Brazilian nature. And here it is really an *other* world that opens up before our eyes, giving us a sense of vertigo: a military fort that is almost lost amid the broad, calm expanse of a river; in the foreground stands a cactus that has as many branches as a tree, a strange animal (it's a capybara, the largest extant rodent); and all around is heat that intensifies the heaviness of the air.

Through the seventeenth-century paintings of Franz Post in Brazil we can still experience the sense of anxiety in discovery, the upheaval caused by the encounter with something undefined, something that does not fit neatly within our expectations. The first thought suggested by the exhibition in the Grand Palais is that the Old World catches the imagery of the New most forcefully when it still does not know precisely what it is dealing with, when information is scarce and incomplete, and it is difficult to separate reality from mistakes and fantasy.

In that same seventeenth century when Dutch painters discovered Brazil, America became an allegorical personage in the works of other artists: it was classified as one of the four parts of the world, and it was accorded a series of attributes like any other mythological figure.

The internal differences within America are recorded in turn in a summary categorization of the various colonies. In order to teach the young Louis XIV geography, he was made to play with geographical-allegorical maps drawn by Stefano Della Bella.

For other painters America offered, almost without mystery any more, a series of stunning views that would enhance the European tradition of landscape painting.

From the eighteenth century America becomes for Europe the embodiment of political and intellectual ideas and myths: Rousseau's noble savage, Montesquieu's democracy, the Romantic fascination with Red Indians, the struggle against slavery.

This allegory corresponds to Europe's need to think of America through its own structures, to make conceptually definable the thing that was and remains the *difference*, perhaps one might say the hard core of America, in other words the fact that it always has something to say to Europe – from Columbus's first arrival there to today – something that Europe does not know.

The allegorical constant is stressed by the final piece in the exhibition, a French painting from the end of the nineteenth century reminding us that the Statue of Liberty was designed and built in Paris between 1871 and 1886. In order to complete the project the restorer of Notre Dame, Viollet-Le-Duc, and the engineer Eiffel, the tower's architect, worked alongside the sculptor Bartholdi. Just as today it stands against the backdrop of skyscrapers, so then it towered above Paris's mansard roofs, before being dismantled and transported to New York by ship.

At this point the exhibition ends, and maybe it could not have gone any further, because in the last hundred years the terms of comparison have changed. There is no longer a Europe that can look down on America from the height of its past, its knowledge and its sensibilities: Europe now contains within itself so much of America – just as America carries within itself so much of Europe – that the interest in looking at each other, which is just as strong and never disappoints – resembles more and more what one feels when looking into a mirror: a mirror that is able to reveal something of the past or the future to us.

[1976]

The Traveller in the Map

The simplest form of geographical map is not the one that seems most natural to us today, namely the map representing the earth's surface as though seen by an extra-terrestrial eye. The earliest need to fix places on a map was linked to travel: it was a reminder of the succession of stops, the outline of a journey. It was thus linear in form, and could only be made using a long scroll. Roman maps were rolls of parchment and we can understand how they were made thanks to a medieval copy which has come down to us, 'Peutinger's Table', which contains the entire road-system of the Empire from Spain to Turkey.

The whole of the known world at that time is on it, in flattened, horizontal form, as in an anamorphosis. Since the important element is the system of land roads, the Mediterranean is reduced to a thin, horizontal, wavy strip, which separates two broader belts, namely Europe and Africa, so that Provence and North Africa are very close, as are Palestine and Anatolia. These continental strips are streaked with lines that are always horizontal, and almost parallel, which are the roads, interspersed with meandering lines, which are rivers. The spaces around are dense with written names and indications of distances; the cities are shown as clusters of little houses of various shapes.

However, these linear types of maps were not restricted to antiquity: there is an English map on a strip from 1675 showing the journey from London to Aberystwyth in Wales, which allows you also to orientate yourself through weather-vanes marked on every segment of road.

On the borderline between cartography and landscape and

perspective painting is an eighteenth-century Japanese roll, over nineteen metres long, representing the whole journey from Tokyo to Kyoto. It provides a detailed landscape where you can see the road climbing over high ground, going through woods, running alongside villages, crossing rivers on little arched bridges, following the gentle ups and downs of the terrain. This is a landscape that is always pleasant to look at, devoid of human figures even though it is full of signs of actual life. (The points of departure and arrival are not marked: the image of the two cities would certainly clash with the uniform harmony of the landscape.) This Japanese roll invites us to identify with the invisible traveller, to follow that road curve after curve, climbing and descending the hills and bridges.

Following a road from beginning to end is particularly satisfying both in literature and in life, and one might ask why in the figurative arts the theme of the journey has not enjoyed similar popularity but instead appears only sporadically. (I now remember that an Italian painter, Mario Rossello, has recently completed a very long painting, also in the form of a roll, representing one kilometre of motorway.)

The need to contain within one image the dimension of time along with that of space is at the origins of cartography. Time as the history of the past: I am thinking of Aztec maps, which are always full of historical and narrative representations, but also of medieval maps such as the illuminated parchment made for the King of France by the famous Majorcan map-maker Cresques Abraham (fourteenth century). And time as the future: like the presence of obstacles one will meet on the journey, and here the weather that is forecast is linked to chronological time; this need is met by climatic maps, like the one drawn up as early as the twelfth century by the Arab geographer Al-Idrisi.

In short, a geographical map, even though it is a static object, presupposes an idea of narrative; it is conceived on the basis of a journey; it is an Odyssey. The most striking example of this is the Aztec codex of Travels [the Boturini codex]. This manuscript uses human figures and geometric outlines to tell the story of the Aztec exodus – which took place between 1100 and 1315 – all the way to the promised land, which is today Mexico City.

(If the Odyssey-map exists, then there has to be an Iliad-map, and in fact all the way from ancient times maps of cities suggest the idea of encirclement, of siege.)

These thoughts came to me while visiting the exhibition on Maps and Images of the Earth at the Pompidou Centre in Paris, and while leafing through the accompanying catalogue.

In an essay in the catalogue François Wahl notes how the representation of the terraqueous globe begins only when the coordinates used to represent the sky are applied to the earth. The celestial parameters (the polar axis, the plane of the equator, meridians and parallels) all meet in the sphere of the earth, in other words at the centre of the universe ('a fertile error if ever there was one', says Wahl). Already Strabo saw geography as a way of bringing the earth closer to the heavens. The roundness of the earth and the grid of coordinates would gain prominence in that they are a projection of the layout of the cosmos on to our microcosm. As Strabo said, 'We have been able to describe the earth only because we have projected the heavens on to it.'

The spheres of the firmament and of our terraqueous globe are put side by side in many Oriental and Western representations. Two gigantic spheres, each 12 metres in circumference – a globe of the earth and one of the sky – are the high-point of the exhibition and occupy the whole of the 'Forum' of the Pompidou Centre. These are the largest globes ever constructed, and were commissioned by Louis XIV from a Franciscan monk from Venice, Vincenzo Coronelli, who was cosmographer to the Serenissima (and author amongst other things of a catalogue of the islands of the Venetian lagoon, with the beautiful title *Isolario*). These globes had been dismantled and placed in chests in Versailles as long ago as 1915: the fact that they have been transported to Paris, restored and remounted on their monumental pedestals and sculpted baroque supports of marble and bronze is on its own enough to make this a truly memorable exhibition.

The heavenly globe represents the firmament as it was on the day of the Sun King's birth with all the figures of the zodiac painted in blue tints. But the great marvel is the earthly globe, in dark brown and ochre colours, studded with figures (showing, for instance, outrages carried out by savages) and inscriptions containing news sent back by explorers and missionaries to fill the voids in those places where the shape of the regions was still uncertain.

California is portrayed by Coronelli as an island, and he comments in a caption: 'Some crazy people say that California is a peninsula . . .' And at another point he says, 'Here people say there is an island, but this is false and I won't put one here.' As for the source of the Nile, after marking it in one place and then moving it after hearing new evidence, Coronelli ends up by inserting a text over the river's flood waters which closes candidly with these words: 'I found I had a space to fill so I inserted this caption.'

All the geographical information on new explorations that arrived in Paris at that time was collected at the Observatoire, where Gian Domenico Cassini kept a huge, flat paper map of the earth up to date. Coronelli was meant to draw his information from that source, which forced him to update his work continuously; but progress in cartography hindered rather than helped this man who still saw geography in the same fanciful way that the ancient compilers had done, rather than as a modern science.

It has to be said that it was only thanks to continuing explorations that the unexplored acquired rights of citizenship on maps. Prior to that, what had not been seen did not exist. The Paris exhibition stresses this aspect of an area of knowledge where every new acquisition opens up the awareness of new lacunae: for instance, in the series of maps where the coasts of South America seen by Magellan in his first voyage were thought to belong to a still unknown Australia. Geography establishes itself as a science through trial and error. (That should please Popper.)

The moral that emerges from the history of cartography is always a lesson about lowering human ambitions. If what was implicit in Roman maps was pride in the identification of the totality of the world with the Empire, later in Fra Mauro's 1459 map we see Europe diminishing compared to the rest of the world. This was one of the first atlases drawn on the basis of accounts by Marco Polo and those who circumnavigated Africa: here the inversion of the cardinal points accentuates this reversal of perspectives.

It is as if representing the world on a limited surface automatically relegated it to a microcosm, hinting at the idea of a larger world containing it. For this reason a map is often situated on the border between two different kinds of geography, the geography of the part and that of the whole, that of the earth and that of the heavens, and the heavens can be an astronomical firmament or the kingdom of God. An Arab tablet made in Constantinople in the sixteenth century bears a very accurate map of the world, surmounted by a (real) compass; a silver pointer pivots on Mecca so the faithful can orientate their prayers in the right direction wherever they happen to be.

From all these aspects we realize that a subjective urge is always present in an operation that seems based on the most neutral objectivity such as cartography. The great cartographic centre in the Renaissance was a city where the dominant spatial theme was uncertainty and variability, since the confines between earth and water there changed constantly: in Venice the maps of the Lagoon had to be updated constantly. (In seventeenth-century Venice Vestri designed a map of the currents which has been shown to be exact in every point by recent satellite photographs taken to determine pollution levels in the Lagoon.) In the seventeenth century the Dutch succeeded the Venetians as the top map-makers, with their dynasties of great artist-cartographers such as the Blaeus from Amsterdam – another place where the confines between land and sea are uncertain.

Cartography as knowledge of the unexplored proceeds at the same rate as map-making that is knowledge of one's own habitat. The origins of this latter kind of map need to be sought in the delimitation of borders in public record maps. One of the first examples of this kind of map is apparently a piece of prehistoric graffiti from the Val Camonica. (It is interesting to note that whereas the borders between properties were scrupulously marked right from the most remote antiquity, similar precision in establishing frontiers between states seems to be only a recent preoccupation. One of the first treaties to fix frontiers in a non-approximate way was that of Campoformio in 1797, in the Napoleonic era, when military and political geography assumed unprecedented importance.)

There is a constant rapport between cartography that looks elsewhere and cartography that concentrates on familiar territory. In the seventeenth century the expansion of the French navy required a regular production of wood, but the forests of France were becoming sparser and barer. Consequently Colbert felt the necessity for a comprehensive relief map of the forests of France, so as to be constantly up to date with the extent of the country's timber resources and to be able to plan rationally the restocking and transportation of wood to the shipyards. It was at that point, precisely to support the navy's expansion, that geographical knowledge of the country's interior became of the utmost importance in France.

Colbert then summoned to Paris Gian Domenico Cassini (1625–1712), a native of Perinaldo near San Remo, and professor at the University of Bologna, in order to run the astronomical Observatory. And here we see once more the link between earth and sky: it was from the Paris Observatory that a dynasty of astronomers, the Cassini family, worked for four generations on a map of France that went into minute detail. The theoretical problems of triangulation and measurement that lay behind the map were at the centre of scientific debate and the very detailed completion of the map would take over sixty years.

The Cassini map, on the scale of one 'line' for every hundred *toises* (1:86,400: a *toise* was about 6.5 feet), is displayed in the exhibition in a reproduction that occupies a whole stand and overflows from the walls on to the floor. Every forest in France is drawn tree by tree, every church has its bell-tower, every village is drawn roof by roof, so that one has the dizzying feeling that beneath one's eyes are all the trees and all the bell-towers and all the roofs of the Kingdom of France. And one cannot help remembering Borges's story about the map of the Chinese Empire which coincided precisely with the physical extent of the Empire.

The human figures which Coronelli felt the need to insert in the expanses of his globe have disappeared from Cassini's map. Yet it is precisely these deserted, uninhabited maps that arouse in our imagination the desire to live inside them, to grow small enough to find one's way amid the dense signs, to run through these maps, to lose oneself in them.

The description of the earth refers on the one hand to the description of the heavens and the cosmos, but on the other it suggests one's own interior geography. Amongst the documents in the exhibition are the photographs of mysterious graffiti which appeared a few years ago on the walls of the new town area of Fez in Morocco. It turned out that they had been put there by an illiterate tramp, a peasant who had left the countryside but had not integrated into urban life and in order to find his way had felt the need to mark the journeys of a secret map of his, which he superimposed on to the topography of the modern city which remained foreign and hostile to him.

The tramp's procedure was symmetrical and opposite to that carried out by an Italian cleric from the beginning of the fourteenth century, Opicino de Canistris. He could not speak, his right arm was paralysed, he had lost most of his memory, and was often in thrall to mystic visions and suffered the anguish of being a sinner, but Opicino had one dominant obsession: interpreting the meaning of geographical maps. He constantly drew the map of the Mediterranean, copying the shape of the coasts all over the place, sometimes superimposing on this drawing the outline of the same map but orientated in a different direction, and he inserted into these geographical outlines drawings of human figures and animals, characters from his own life and theological allegories, sexual penetrations and angelic apparitions, placing alongside them a dense written commentary on the story of his misfortunes and prophecies concerning the destiny of the world.

In an extraordinary example of 'art brut' and cartographic madness, Opicino simply projects his own interior world on to the map of lands and seas. Using an inverse procedure, the Society of 'Précieuses' in the seventeenth century would try to represent psychology using the code of geographical maps: the map of 'tenderness' devised by Mlle Madeleine de Scudéry shows the Lake of Indifference, the Rock of Ambition and so on. This topographical, horizontal idea of psychology, which shows relationships of distance and perspective between passions that are projected on to a uniform expanse will later give way to Freud and his geological and vertical idea of depth psychology, made up of superimposed layers.

[1980]

The Museum of Wax Monsters

In a window looking out on to the street a young woman lies supine in a white, flowing dress adorned with lace frills, her sleeping face with its delicate features the colour of mortuary yellow, her chastely covered breast rising and falling with regular breaths. A little bit further on a poster shows a colour photograph of Siamese twins, or rather one single male child who divides above the stomach into two identical boys. Around all this is a canvas façade painted red with gilded adornments and the words: *Dr P. Spitzner's Great Anatomical and Ethnological Museum*.

For over eighty years, from 1856, Dr Spitzner's anatomical wax museum was a fairground attraction, especially in the towns of Belgium. Initially it had been set up in Paris, with the full endorsement of a scientific institution (eighty of its exhibits came from Dr Dupuytren's famous collection of pathological models); but various vicissitudes turned it into an itinerant museum which found its proper place amidst fairground stalls, merry-go-rounds, shooting galleries and menageries. All the while it proclaimed its educative and moralistic intentions: the foreword to its guide opened with a kind of ten commandments for a healthy life, which is the first joy and duty of good citizens. The horrific visions that the museum displayed (tumours and ulcers and buboes, or livers with cirrhosis and stomachs with fibrosis) were meant to inculcate in the young the terrors of venereal diseases and alcoholism. However, the sections devoted to these 'culpable' diseases was just a part, albeit an important one, of the exhibition, which as a whole seemed to invite onlookers to fix their eyes on things that we are usually inclined to turn them away from: the possible deformations of our flesh, the hidden physiognomy of our innards, the agony we feel within ourselves if we see a surgical operation.

In addition to this schooling in horror there was also, strangely, ethnological documentation: a parade of wax statues representing bushmen or Australian or American Indian savages, life-size, a sight which in those pre-cinema days must have been much more dramatic than we can imagine today. On closer inspection, the motif that is common to the whole museum dominated also in this ethnological section: a nakedness that was 'different', intimate like all nakedness but distanced by disease, deformity or the 'otherness' of another civilization or race, with in addition the unease that wax arouses in us when it imitates the pallor of human skin.

Who this Dr Spitzner was in real life is not clear. One suspects that he was not a doctor at all. In the photographs he and his wife have more the look of fairground impresarios than apostles of science; but one can never tell. Certainly, the sadism which is an essential component of the visual world he offers us was of a different order from the more poetic sadism of the Florentine Clemente Susini, or the more wizard-like version of the Neapolitan Raimondo di Sangro, or the purely spectacular sadism of Marie Tussaud, who was English by adoption. But these last three names all belong to the eighteenth century, with all the complexity of intellectual and psychological attitudes that characterized that period; whereas the date of the foundation of the Spitzner Museum takes us right into the age of positivism and scientism and popularizing pedagogy; a date that is no less glorious, however, if one thinks that it is the same year as the publication of *Les Fleurs du mal* and *Madame Bovary*, and of the related court cases against what was then abhorred or revered as an 'exploration of the truth'.

As in those lofty cases, so also the not easily definable enterprise of Dr Spitzner had to struggle against the hostility of the prudish, censorship by the authorities, the protests of fathers of families; and the same battles were refought in our own century when Mrs Spitzner, after being widowed, started up the travelling museum again in the 1920s. The fact is that, in the memories of various Belgian writers and artists, their first terrified entry into Dr Spitzner's pavilion occupies a powerful place: suffice to say that the artist Paul Delvaux declared that this was the fundamental experience in the formation of his visionary world, even before his discovery of De Chirico.

The museum went missing during the war (the exterior billboards, certainly not a negligible part of its fascination, were destroyed in a bombing raid), but was rediscovered in a warehouse, and now Dr Spitzner's Museum has been reconstructed and put on temporary display by the Belgian Cultural Centre in Paris, in the Place Beaubourg. The first thing that strikes you is how the faithful imitation of nature, instead of seeming timeless, is full of the colour of that period. It is the look with which these models have been conceived that is nineteenth-century: a mixture of attraction and distance at the same time, of celebration and condemnation of the 'truth'.

In the reconstruction of its environment they have tried to preserve that atmosphere that lies somewhere between the scientific and the seedy, an atmosphere that is that of the hospital laboratory, the morgue and the fairground booth (all of which it must have had at the time), including the penumbra against which the cadaverous nudes stand out and the muffled fairground music that sounds as if it is being played by a country band. All that is missing are the shouts of the touts and the guides who – according to the chronicles of the time – would demonstrate the 'Anatomical Venus', which could be dismantled into forty pieces, moving from the seductive fragrance of her skin to the dark tangle of blood-vessels and ganglia, to the web of nerves, and the whiteness of her skeleton.

Not just wax models but also natural exhibits are on display, such as for instance a complete human skin, that of a thirty-five-year-old man (a unique piece, the catalogue warns us, as no museum holds anything like it): this human carpet, which is squashed like a flower inside the pages of a book, seemed to me the most friendly and comforting thing in the midst of everything else. I have to admit I have never felt any attraction for innards (just as I have never felt any strong urge to explore psychological depths); that perhaps explains my preference for this man who is completely extended, his whole surface unfolded before us, devoid of any thickness or hidden intention.

All in all, apart from a few notes about its atmosphere, I cannot really be a good chronicler of the Spitzner exhibition: my gaze tended instinctively to avoid any image in which insides spilled outwards. I preferred not to loiter, especially in the pavilion devoted to venereal diseases, comforted by the cheering news that some clinical aspects on show there have disappeared today thanks to medical advances. (This is said in the catalogue, which boasts that even the medical specialist will find the exhibition of historical interest, since certain lesions caused by syphilis have now 'abandoned the pathological scene'.)

I prefer instead to lean in contemplation over the glass bell-jar containing a reproduction of the guillotined head of the anarchist Caserio, a wax model made immediately after the original head fell into the basket (1894). His sliced neck is as fresh as meat in a butcher's shop, his expression is fixed for ever with staring, rolled-back eyes, dilated nostrils, locked jaws: the effect is not dissimilar to that produced by a sudden flash photograph, but here the objectification is total, without any trace of subjective framing.

The most incredible example of sadist-surrealist fantasy is to be found among the representations of the various phases of childbirth and gynaecological operations. A complete model of a patient undergoing a Caesarean section lies with eyes wide open, her face distorted by pain, her hair impeccable, her calves tied together, dressed in a long, lace gown, which is open only at the part of her body which has been cut open by the scalpel, where the baby appears. Four male hands are placed on her body (two operating, and two holding her waist): fine wax hands with manicured nails, ghostly hands since they are not supported by arms but adorned only with white cuffs and with the ends of the sleeves of a black jacket, as though the whole ceremony was being performed by people in evening dress.

One of the attractions that brought (and still brings) visitors flocking to the exhibition was the *Gallery of Freaks*. There is a wax facsimile of the private parts of a certain John Chiffort, 'born in the county of Lancaster, and reproduced from life when he was twenty years old; he possesses three legs and two penises, both capable of reproduction'. Were it not for his central leg, which has atrophied and is frankly very unpleasant to see, the two penises, which are symmetrical and parallel to each other, have such a natural, gracious look that you could easily believe it might be normal for all males to be so endowed.

The opposite case to this is that of the Tocci brothers. Born in Sardinia in 1877, each of them possessed his own head, and his own perfectly normal pair of arms and shoulders, but from the level of the stomach downwards they were one single person, with just one stomach and a single pair of legs. Their wax model (which is reproduced also in the posters for the exhibition) shows them apparently at the age of nine or ten, and the emotion they arouse is heightened by the fact that their faces are those of two very handsome boys with a lively air about them. 'They currently enjoy excellent health and have been on tour in the main European capitals. Without a shadow of a doubt they constitute the most curious phenomenon that has ever been seen.' To these words from the old catalogue is added a more recent note: 'In 1897, after making their fortune, the Tocci brothers married two sisters and retired to a property near Venice, where they would die in 1940 at the age of 63.'

The problem is that this information quoted in the catalogue is largely untrue. I can confirm this because in these last few days I have come across the recent volume entitled *Freaks*, by Leslie Fiedler, which, apart from chapters on dwarves, giants, bearded women and hermaphrodites, contains about thirty pages on Siamese twins which are full of essential information. From this source it turns out that Giovanni Battista and Giacomo Tocci, who were baptized as two separate people even though from the seventh rib downwards they were just one person, had to put up with another severe handicap: their single pair of legs was unable to support them or to walk. (In fact in Dr Spitzner's wax model we see them leaning on a railing.) This immobility severely limited their possibilities in exhibitions of 'living phenomena', and as a result, after a rather brief but exhausting international tour, they were forced to give up their circus career and retired to Italy, where they sadly died (I can't find the date, but presumably at a young age).

The news about the marriage with two sisters probably derives from the fact that the account has been contaminated with another story, a true one (the only one of its kind that can be considered as having in some sense a 'happy ending'). This was the story of the eponymous Siamese twins (in other words the twins whose fame is the reason we call 'Siamese' all twins who have one part of their body attached to the other twin). Chang and Eng were born in 1811 in Siam into a poor Chinese family and died in the United States in 1874. They quickly fell into the hands of unscrupulous impresarios, who transported them to America, thinking they could use them as their own goods and chattels, but Chang and Eng were able to become independent and to manage their own fortune without being exploited even by the grasping Barnum, in whose circus they appeared until 1839.

The story of Chang and Eng represented the triumph of both Chinese shrewdness and the American belief in overcoming adversities and prejudices: in fact they managed to retire to the North Carolina countryside and to gain the respect of the closed world of white farmers, so much so that they married two sisters, daughters of a wealthy landowner who was also a pastor in the Baptist Church. With their wives they had twelve and ten children respectively, all of them healthy, so their descendants nowadays amount to a thousand or so American citizens.

The image of the Tocci brothers on the wall-posters struck the imagination of Mark Twain, who drafted a story inspired by their case, just as the fortunes of Chang and Eng provided him with material for another story. (The theme of the 'double' is a recurrent motif in his oeuvre.) Fiedler's book, whose subtitle is *Myths and Images of the Secret Self*, records and blends historical facts with literary and cinematographic inventions and with evocations of mythical archetypes. The most interesting pages of the book are the true stories: the lives of 'living phenomena' in the world of the circus, almost all of them very sad tales.

However, the starting-point for this volume by Fiedler is a reflection on the alternating cultural fortunes of the term *freaks*, which at one time was associated with fascination and horror, and which now has been appropriated 'as an honorific title by the kind of physiologically normal but dissident young people who ... are otherwise known as "hippies", "longhairs", and "heads"' (Leslie Fiedler, *Freaks. Myths and Images of the Secret Self*, New York: Simon and Schuster, 1978, p. 14). From this premise Fiedler sets out to conduct research into the value that forms of physical 'diversity' have had in various cultures, as an examination of the confines and roles that define human existence. Seen from this perspective, Dr Spitzner's waxwork museum may offer food for further thought.

[1980]

The Dragon Tradition

Like so many other things, the study of dialects in France began in the Napoleonic era. In 1807, the office of statistics in the Ministry of the Interior launched an inquiry throughout all the Prefectures of the country. The task was to put together a collection of versions of the parable of the prodigal son in the different patois and dialects spoken in France. They decided on this base-text after considering other alternatives (for instance a collection of Sunday sermons) and in the end they opted for that particular episode since, as recounted in the verses of St Luke's Gospel, it offered the prerequisites of simplicity and universality as well as a typical, everyday lexis. When the Restoration came, the office of statistics was suppressed, but the research was continued by the Société royale des Antiquaires until they had assembled 300 versions.

These details alone are enough to give an idea of how questions concerning the study of popular culture in France are framed differently from in Italy. In France the multiplicity of local cultures lies almost hidden behind the massive hegemony of the linguistic and cultural unity of the nation (whereas in Italy these proportions are inverted), and the urge to gain knowledge of this world starts with the realization that they are in the process of disappearing. (What is different in France and in Italy is the pace of historical survival: in my own country it seemed as if up to yesterday traditional customs and mentalities were ineradicable, but then they suddenly disappeared overnight, whereas in France they quickly became marginalized but as such they have enjoyed a very lengthy period of survival.)

The exhibition mounted at the Grand Palais, entitled Yesterday for Tomorrow: Trades, Traditions and the Nation's Legacy, traces the origins of the 'ethnographical' discovery of France in the period of the Enlightenment, when the *Encyclopédie* valorized and catalogued the tools and operations of the 'mechanical arts'. Alongside the stalls dedicated to artisan trades the exhibition features an authentic loom from the period, a loom for making stockings, along with beautiful embroidered silk stockings also from the eighteenth century. 'The loom for making stockings is one of the most complicated and rigorously logical machines that we have,' is the philosophical comment offered by Diderot. 'One could consider it as just one single piece of reasoning whose conclusion is the making of the stockings.'

At the same time, a decisive step was taken by the Société royale d'Agriculture, when the shepherd, a figure who had been made sugary sweet by the pastoral convention in art and literature, became a subject of technical knowledge with manuals such as *A Treatise on Wool-Bearing Animals, or the Method for the Rearing and Managing of Flocks* (1770) or *Instructions for Shepherds and Owners of Flocks* (1782). In this case too the Grand Palais exhibition places beside written and artistic documents the objects and tools they actually used: in this case, dog-collars with iron spikes or crooks with spoon-shaped tips for throwing clods of earth at disobedient rams.

The doctors of the Société royale de Médecine were ethnographers without realizing it: they combed the countryside to trace the origin and spread of epidemics and occupational diseases. Their *Medical Topographies* contain descriptions of the living environments of rural families, and also of the work involved in skills such as lace-making and glass-blowing.

The French Revolution pushed the people to the forefront of history, but did not do much to get to know them in concrete terms, despite the efforts of a curious kind of scholar, the Abbé Grégoire, who decided to use the network of local 'patriotic societies' to distribute a questionnaire on the dialects and customs of the countryside. The fact is that the Revolution was forced to acknowledge the cultural divide separating the city populace (the *sans-culottes*, armed with pikes) from that of the countryside (the Royalist *Chouans*, armed with sickles), and relegated the 'savage side of France' to the vestiges of the past to be eradicated without quarter. In short, as far as the two aspects of Enlightenment culture are concerned – that of linear, unifying progress, and that of a detailed knowledge of the country's diversities and the rationale behind them – the Revolution appropriated only the first aspect, consistent with the logic of Jacobin centralization.

The revival would come about in the climate of the Romantic reaction to the Enlightenment, when the 'Celtic Academy' resolved to reconstruct the image of an autochthonous civilization, that of Druidic Gaul, in opposition to the Graeco-Roman model so beloved by the revolutionaries. But these ideological contrasts belong more to our schematic way of looking at things than to reality: for the scholars of the Celtic Academy were still people who had been educated in Enlightenment culture, and their inquiries and research methods were models of scientific modernity.

One of the first initiatives promulgated by the Celtic Academy was a census of dragons: in about twenty French cities a papier-mâché dragon was (or is still) carried in a procession once a year. The legend behind the festival is more or less the same everywhere: maidens being offered to the monster, then their rescue by a male or female saint. The dragon has two aspects: a terrifying enemy in the legend, it becomes in the procession a carnivalesque, almost benign presence, with which the city identifies and from which it seeks protection. The readers of *Tartarin*, who remember how the town of Tarascon was proud of its Tarasque dragon, will find in this exhibition a huge naive painting of the people of Tarascon carrying their 'Tarasque' through the streets.

Reproduced in the exhibition posters, this painting promises the visitor a more lively and jovial set of exhibits than is in fact encountered. As we can easily imagine, nothing is more boring than walls full of illustrations of life in the countryside in the nineteenth century. And the photographs of washer-women in Breton costume are no more exciting. The fact is that in nineteenth-century art and literature there was a comforting ideology behind this image of rural life: the countryside was the healthy world of lost virtues as opposed to the town. From such a false and tedious idea could only come false and tedious representations, as the exhibition abundantly proves.

What is interesting, on the other hand, are the exceptions to this picture. For instance, Maurice Sand, son of George Sand (who today is coming back into fashion, being reprinted and read both from a feminist standpoint and in precisely this 'ethnographic' way), illustrated his mother's 'rustic legends' with Romantic drawings à la Doré, in a visionary, hallucinatory style, which can also be quite disturbing. And above all there is an ethnographic draughtsman, Gaston Vuillier (1845–1915), who was fascinated by practices of witchcraft and occultism in the countryside, and who possessed both a scrupulous documentary fidelity and a sense of unusual effects. (It turns out that he also visited Sicily and Sardinia, making drawings of magical practices there: it would be worthwhile tracking these down.)

Naturally, it is the objects that speak more than the paintings and photographs and the usual, predictable regional costumes. A large amount of this material comes from the Museum of Popular Crafts and Traditions, which for the last twelve years has been based in the Bois de Boulogne in a site that is a model of museum display. While in its home museum the material is presented using a systematic classification, here in the exhibition (which has been organized by the museum conservator himself, Jean Cuisenier) a historical approach prevails: it presents the history of France's 'ethnographic' interest in itself. As a result, going through the exhibition is a bit like leafing through the *History of Folklore in Europe* by our own Giuseppe Cocchiara, a book written about thirty years ago but which remains the most useful brief account for contextualizing in historical terms the various types of approach adopted by high culture in studying the areas that are most remote from itself.

In recent times, historical research has been trying to extend this network of relationships backwards in time, in other words to define oppositions and interactions between high culture and popular culture right from the time of the Renaissance, if not indeed from the Middle Ages. This is the direction taken by the most recent work on the subject to come out in Italy, a book by the English historian Peter

Burke: *Popular Culture in Early Modern Europe* (New York: Harper and Row, 1978).

Naturally, also from this perspective, the cusp of the eighteenth and nineteenth centuries remains a crucial period. 'The discovery of popular culture,' writes Peter Burke,

> took place in the main in what might be called the cultural periphery of Europe as a whole and of different countries within it. Italy, France and England had long had national literatures and a literary language. Their intellectuals were becoming cut off from folksongs and folktales in a way that Russians, say, or Swedes were not . . . It is not surprising to find that in Britain it was the Scots rather than the English who rediscovered popular culture, or that the folksong movement came late to France and was pioneered by a Breton, Villemarqué, whose collection, *Barzaz Braiz*, was published in 1839. Again, Villemarqué's equivalent in Italy, Tommaseo, came from Dalmatia, and when Italian folklore was first studied seriously, in the later nineteenth century, the most important contributions were made in Sicily . . . In Germany too the initiative came from the periphery; Herder and Von Arnim were born east of the Elbe. (pp. 13–14)

The English historian's thesis is confirmed by the dates of the documents in the Paris exhibition: one could say that France was the last nation in Europe to start studying its own popular and rural traditions, so much so that the monumental work in this area, *The Manual of Contemporary French Folklore* by Arnold Van Gennep, came out (in nine volumes, but incomplete) only in the twentieth century, between 1937 and 1958. But for me the important point is something different: it is always the awareness of something that is about to be lost that brings about the *pietas* for these humble vestiges. The 'centre' only comes to this awareness later on, when its drive towards cultural homogenization is, one might say, complete and there is not much left to salvage; the 'periphery' notices this beforehand, seeing it as a threat that comes from the pressure to centralize.

This year is the Year of Heritage in France, and the exhibition is organized in this context, with special attention being paid to the role played first by private collections and the antiques market in valorizing rustic ceramics and wooden sculptures, subsequently by regional museums, and now by the 'regional parks' which are planning a wider campaign to safeguard the environment. The word *patrimoine* ('heritage, tradition'), which is dear to the old heart of Balzac's France, of a country that 'saves', creates an impression of something solid and substantial, something that can be turned into capital (whereas we Italians talk of *beni culturali*, 'cultural goods', an expression devoid of any connotation of possession or concreteness). Perhaps only this echo of material interest can counterbalance the instinctive gesture of contemporary man: that of throwing things away.

[1980]

Before the Alphabet

Writing first emerges in Lower Mesopotamia, in the land of the Sumerians, whose capital was Uruk, around 3300 BC. This was the country of clay: administrative documents, bills of sale, religious texts or those glorifying kings were engraved with the triangular point of a reed or quill on clay tablets which were then dried in the sun or baked. The surface and the instrument ensured that primitive pictograms quickly became simplified and stylized in the extreme. Pictographic signs (for a fish, a bird, a horse's head) lost their curves since these did not come out well on clay: in this way the resemblance between the sign and the thing represented tended to disappear; the signs that dominated were those that could be drawn with a series of instant strokes of the quill.

In general these signs had a triangular apex that then prolonged itself into a line forming a kind of nail shape, or forking into two lines like a wedge: this was cuneiform writing (*cuneus* means wedge in Latin), writing that transmitted an impression of rapidity, movement, elegance and compositional regularity. Whereas in inscriptions sculpted on stone the predominant sequence of signs was vertical, writing on clay tended naturally to extend along horizontal, parallel lines. This linear, tense, sharp movement of the pen which we recognize in cuneiform documents would become the movement made by anyone who uses a fountain pen or biro in our own times.

From that point onwards, writing would mean writing quickly. The real history of writing is the history of cursive handwriting; or, at least we could say that it is to the cursive form of writing that cuneiform owes its fame. Economy of time, but also economy in terms of space: to place as much writing as possible on a surface was a tour de force that was quickly embraced. One tablet fragment that has been preserved is two centimetres by two, and has thirty lines of liturgical lamentations in a microscopic cuneiform.

The Sumerian language was agglutinative: monosyllables were accompanied by prefixes and suffixes. As the signs became detached from the pictograms and ideograms that they originated from, they gradually became associated with syllabic sounds. But cuneiform writing continued to retain vestiges of the various phases of its evolution. In the same text, in the same line even, ideogram signs (the king, the god, adjectives like 'shining', 'powerful') are followed by syllabic phonetic signs (especially for proper names: the priest Dudu is written with a drawing of two feet since 'Du' means 'foot') and grammatical-determinative signs (for the feminine they use a triangular sign which originally was the female pubes).

The Louvre holds quite a number of documents of this type – clay tablets, engraved stones or metal plaques – but it was only the specialists that could make them speak to us. Now the exhibition that has just opened at the Grand Palais, dedicated to The Birth of Writing: Cuneiform and Hieroglyphics, has put on display over 300 items (almost all of them from the Louvre, and one or two also from the British Museum), accompanied by extensive and informative captions. This whole exhibition is one that needs to be read: both the explanation panels, which are indispensable, and the writing in the original documents, on stone, clay or papyrus, however little of these one actually manages to read. Perhaps there is slightly too much material, in terms both of exhibits and of information; but the visitor whose eyesight is not too strained and who overcomes the fear of becoming overloaded (a phase which is inevitable at the outset), can in the end claim to have understood how we reached alphabetic writing.

The linearity of writing has a history that is anything but linear, but it is played out entirely in a strictly limited geographical area, over the course of two and a half millennia: everything takes place between the Persian Gulf, the Eastern Mediterranean coast and the Nile (Egypt represents a lengthy chapter in itself in this history). If it is true that Indian writing too and probably even Chinese script derive from the same stock, we can conclude that for writing (as opposed to language) one can talk of a single birth. (And pre-Columbian America? The exhibition does not broach this problem.)

What is certain is that writing (unlike language) is a fact of culture not of nature, and that in origin it concerns a limited number of civilizations. This is mentioned in the catalogue by Jean Bottéro (famous for a brilliant essay on divination techniques in Mesopotamia, in the volume edited by Vernant, *Divination et rationalité (Divination and Rationality)*). He points out that the vast majority of spoken languages have never been written languages, even though in our own times many of them have ended up undergoing alphabetization by another culture.

Why Lower Mesopotamia precisely? Five thousand years ago, in those arid lands a new political and economic system was formed which had at its centre the city and a priestly monarchy; irrigation works made possible huge agricultural development as well as a demographic explosion. This led to the necessity for a complicated system of accounting to check the taxes being raised, exchanges of goods and land registries for a huge number of people over vast expanses of land. Even before being used for writing, clay, which was an essential memory aid, was used to set down data that were solely numerical; then suddenly alongside the notches corresponding to figures they began to carve signs representing goods (animals, vegetables, objects) or people's names.

Are we to conclude then that it was merely a practical, mercantile or indeed tax-related necessity that opened up the boundless spiritual realms of written culture? The story is a bit more complex than that. The earliest forms of graphic symbols were adopted in memoranda regarding income and outgoings because these forms had already been developed in the world of art, especially on painted ceramic vases. Already for some time now, on funerary or cult objects as well as on objects in daily use, the 'name' of the individuals or the gods had been represented by shapes that were at the same time expressions of admiration, fear, love or domination: states of mind, attitudes towards the world. The expression of something that we can already define as poetic on the one hand and economic records on the other are thus the two needs that preside over the birth of script; we cannot write the history of writing without keeping in mind both elements.

Towards the middle of the third millennium BC, cuneiform writing passed from the Sumerians to the Akkadians (their capital was Agade or Akkad), who spread the practice throughout their Empire as far as Northern Mesopotamia. The Akkadians had a Semitic language (with a tri-consonantal root) which was completely different from Sumerian. The same signs were used to designate the same things, even though they corresponded to sequences of different sounds (in other words, having been phonetic, they went back to being ideogram-based) or they were identified with the new sound, losing the link with the former meaning (in other words, having been ideograms, they became phonetic).

Everything became more complicated thanks also to the proliferation of signs (several hundred of them); and yet it was through the Akkadians that cuneiform writing spread throughout the whole of the Middle East (we find it again in the library at Ebla, which was recently discovered), passing to the Assyro-Babylonians and the Syrians, to the Elamites of Southern Persia, to the Canaanites of Palestine, to the Aramaic peoples whose language spread from India to Egypt in the first millennium BC.

If the most ancient documents provide individual words, mostly names, which are not articulated into sentences – as though men had learned *how* to write before knowing *what* to write – by the time of Nineveh and Babylon these thick tangles of hens' footprints can give us the epic of Gilgamesh, or provide us with a dictionary, a library catalogue or a treatise on the measurements of the Tower of Babel (apparently it was a seven-storey ziggurat, 90 metres high).

Whereas in Mesopotamia one can follow the evolution from pre-writing (or pre-numbering) to cuneiform script, in Egypt hieroglyphics suddenly appear, admittedly somewhat tentative and disorganized in the beginning, but without any antecedents that we know of. Does this mean that writing was imported into Egypt from

Mesopotamia? Chronology would seem to confirm this hypothesis (only a couple of centuries separate the first pictograms in Uruk and the first hieroglyphics), but the Egyptian system is totally different. Are we then dealing with an autonomous invention? Perhaps the truth lies somewhere in between: the Egyptians had close trading links with Mesopotamia and soon learned that the Sumerians 'wrote'; this news opened up new horizons for their inventive capacities and before long they elaborated a highly original way of writing that would be solely theirs. Already by 3080 BC, there are about seventy funerary steles which prove that Egyptian hieroglyphics contained no fewer than twenty-one alphabetical signs (our consonants were all already present), plus other signs which designated groups of letters, rebus-words and signs which served to determine the sense in which other signs were to be understood.

The oscillation between figurative elements and writing haunts graphic activity for at least two millennia, and it is this ambiguity that makes the exhibition at the Grand Palais beautiful to see as well as instructive to 'read' and explore. An Egyptian stone stele of great beauty has a bas-relief of a falcon, a serpent and the walls of a city; you could say anything about this harmonious figurative composition apart from that it makes you think of something written; and yet the walled city is the sign that designates a king; the thoughtful bird is the god Horus, whose terrestrial form is that of a king, and the sinuous snake stands for the name of the king. On the other hand, some reliefs of birds are nothing but the graphic models for U and A made by a sophisticated designer from the Ptolemaic age.

Even when hieroglyphics have become a well-codified writing system, the Egyptian scribe prefers not to follow a linear arrangement, but to compose groupings that are aimed at the beauty of the whole arrangement, even though this goes against the logical order and proportions between the signs.

The arrangement in vertical columns, which had prevailed before the horizontal mode (from right to left) triumphed (during the Middle Empire), allows one the freedom to read the hieroglyphics in a vertical or horizontal (starting from the right or from the left) sequence: that was when the scribes' erudite games began, as they combined one direction for reading with another and invented the crossword!

In the same period there were hieroglyphic statues, or all-round rebuses: a dense sculpture of the eighteenth dynasty compacts together into a single block a serpent, two raised arms, a basket, a kneeling woman: what can this mean? The cryptographic explanation (which I shall not try to summarize) leads to a meaning not through the logic of images but through a succession of sounds.

In the reliefs and pictures on tombs in ancient Egypt, characters represented by figures are placed alongside columns of writing which are the characters' words, just as in today's comics. But the great thing is that these human figures, stylized and all seen in profile, seem to have the same nature as the graphic signs, from which they differ only in size, whereas words in hieroglyphics still belong to the world of figures. The analogy between comics and these Egyptian procedures is brought home by the Grand Palais exhibition, for it has commissioned comic book artists to create modern equivalents of these scenes where Pharaohs and priests exchange hieratic phrases with each other, warriors roar threats and insults, and sailors and fishermen address each other with mocking remarks.

The universe of images is infinite, so new signs could always be added to the galaxy of hieroglyphics: 'Ptolemaic' writing managed to reach a total of more than 5,000 signs. In this elasticity lay the practical inconvenience of hieroglyphic writing but also its poetic richness: in a funerary papyrus the name of the god Amun was written in five different ways, each one of them corresponding to a different philosophical and religious content. But this inability to become a closed system prevented hieroglyphics from expanding beyond Egypt, whereas cuneiform writing had conquered the whole of the Middle East.

However, in the first millennium BC Egyptian scribes developed a cursive writing of their own, a system that was even more rapid and expressive than cuneiform, and which would last until the first centuries of our era. The owl, in other words the letter M, first became a scrawl, then a kind of Z, then a kind of 3, but all the while retaining something of its initial hieroglyphic. Here too (as with cuneiform) the writing instrument and the material were crucial: in this case ink and papyrus. Everything was ready: there was nothing more to add to the art of writing. Except one thing: the alphabet.

The alphabet, or rather the series of signs where each one corresponds to a sound and when grouped in various ways can represent all the phonemes of a language, started with twenty-two signs on the Phoenician coast (Lebanon today) around 1100 BC. Mohabite, Aramaic, Hebrew and later Greek all derived directly from 'Phoenician consonantic linear'. Coptic, which developed from the Egyptian cursive, and Arabic alphabets have a separate history, but still connected with this one.

Careful: I notice that now the specialists write 'Phoenicians' in inverted commas, or they say 'Those peoples known as the Phoenicians . . .'. I don't know what is behind this; and I can tell you I am in no hurry to find out. The story of the Phoenicians was one of the few certainties I still retained. Now, although it seems confirmed that they invented the alphabet, it appears that a suspicion has emerged that they never existed. We live in an age when nothing is sacred, nothing and nobody.

[1982]

The Wonders of the Popular Press

A polar bear devouring a young girl stands out from the posters advertising an exhibition dedicated to human interest stories in the press (Le Fait divers, an exhibition in the Museum of Popular Traditions and Crafts, Paris). These extraordinary events that catch the popular imagination are presented not from the point of view of the history of journalism but as a modern form of folklore.

The polar bear comes from an illustration in *Le Petit Journal* from 1893 depicting what it calls 'The Frankfurt Suicide'. An unusual suicide, described briefly by the press but with sadistic details that were bound to hit home: a young girl in domestic service, desperately unhappy in love, goes to the zoo, takes her clothes off and, while singing a song, enters the cage of an animal which then leaps on her.

A considerable amount of the material exhibited comes from the illustrated supplement to *Le Petit Journal*, which with its colour plates would become the model for Italy's *Domenica del Corriere* (but about thirty years before: *Le Petit Journal* starts in 1863, the *Domenica* in 1899). In it we see tigers and elephants escaping from circuses, Dantesque tragedies in the Paris sewers, a crime of passion in a butcher's shop, a suicide inside a tomb, another suicide carried out using a home-made guillotine, a naked man wearing only a top hat and side-whiskers entering a shop while the women cover their eyes. It has the distinction of being the first paper to illustrate the news (even though it is all reconstructed through the illustrator's imagination), thus anticipating documentary films and television, but it also holds the record in linguistic and above all conceptual terms, since the term 'fait divers' appears for the first time in *Le Petit Journal*.

The period covered by the exhibition, however, goes back much further, starting with the printed sheets with crude engravings and rudimentary texts that were sold in markets in the eighteenth century, with stories and images of bandits and crimes. These continued throughout the nineteenth century, going by the name of *canard*. The term *canard*, literally 'duck', but meaning an 'unlikely, probably false story', has been present for centuries in popular French, and nobody knows its origin: some say that those who sold these *canards* in the fairs announced their presence with the sound of a little horn which resembled a duck's quack, but this etymology has not been proven. Broadsheets similar to the *canards* of the eighteenth and nineteenth centuries continued down to our own times, still using printing technology that had not progressed very much, and circulating verses of songs about current affairs. For instance, in 1909, the Messina earthquake was depicted with Roman temples in ruins crushing the populace.

The character who dominates such documents is of course the law-breaker: brigands in the countryside and, starting in the nineteenth century, perpetrators of organized crime in the city, but also individual murderers, killing for money, for passion or out of insanity. We learn that the word *chauffeur*, which for us evokes dynamic images of cars at the start of the century, had between the eighteenth and nineteenth centuries a terrifying meaning: *chauffeurs* was the name given to the brigands who attacked houses in the countryside and would burn the feet of their victims to force them to reveal where they had hidden their money.

The fascination which the outlaw and the criminal exercised on the imagination (in an age when crime had not yet become an industry like any other) is proven also by illustrated postcards portraying famous bandits and assassins: the famous bloodbath between members of the Parisian Apache gang over the beautiful eyes of the blonde known as 'Casque d'Or' (which would later inspire a fine film in the early 1950s) is portrayed as though it were a photo-story magazine in a series of postcards from 1907. Similarly in 1913 the images of the Bonnot Gang also ended up on illustrated postcards.

It is not only the cruelty of the crime that excited curiosity, but also, from the very start, its counterpart, the cruelty of the punishment.

The guillotine is a major motif in popular iconography (and in songs); a series of illustrated postcards that have the objectivity of gloomy, black-and-white photographs has preserved for us a wide-angle shot of the prison, an overview of the apparatus, details of the wooden semi-circle and the basket, and even a framed shot of the garage where the contraption was kept when not in use: the bureaucratic and technological character of the start of the century is documented here in its most depressing aspect.

The old executioners' custom of selling the rope from a hanging as an amulet continues in a macabre cult of the relics of guillotine executions. Here one can see displayed, framed and beneath glass, a collection containing the collars from Caserio's jersey and shirt which had been cut off for his final *toilette* before his execution: Caserio was the anarchist responsible for the fatal attack on President Carnot (1894). (The film *Danton* by Wajda, which is currently being shown in Paris, dwells on the details of this *toilette* during the French Revolution.)

Murder, like sainthood, produces relics: the furniture in Landru's house was put up for auction in 1923, and of course the highest price paid was for the wood-burning cooker where Landru got rid of the remains of his 'fiancées'. Here we are told that 'an Italian paid 40,000 lire for it'. (Is it still in Italy? Should it be considered part of the country's heritage to be protected?)

The trial is the moment when the evocation of the bloodshed and that of its punishment are present at the same time, and it is by starting with the trial itself that journalism arouses people's emotions. It is no accident that many of the documents here revolve around 'famous trials'. From as early as 1825, with the birth of *La Gazette des Tribunaux (The Court Gazette)*, such cases could count on a specialized area of journalism, which would in turn inspire great writers, from Stendhal and Balzac to Eugène Sue, as well as writers of serial novels.

Black humour about crimes and executions was current not just amongst the more blasé spirits, but also in the popular press: in 1884 we find an *Assassins' Journal*, registered as 'the official organ of Knifers United' ('Meeting time and place for subscribers: midnight, at street corners'). But I do not know whether it ever went beyond its first issue.

'Bloodstained hotels', where the owners murdered their clients in their sleep and burned them in the stove, form another topos which crime journalism from nineteenth-century rural France would pass on to literature and theatre (the most recent example being Camus's *Le Malentendu* (*The Misunderstanding*)). The most notorious such hotel was that of Peyrebeille, where M. and Mme. Martin, along with their servant Rochette, also known as 'The Mulatto', eliminated a number of people whose total was never established exactly. They were then guillotined in 1833 on the very site of their crimes. That was all that was needed for their hotel to turn into a tourist attraction, with postcards and souvenirs.

These violent stories provided the mythical raw material that was then taken up by popular literature (which closely followed such crime reporting, with instalments selling at 10 centimes each on famous crimes that had passed into literature). It was also taken up by the theatre, which specialized in these crimes and got its macabre inspiration from the name of the Boulevard on which it was situated, Boulevard du Crime (immortalized in Carné's film *Les Enfants du Paradis*). Other areas that derived inspiration from these stories were the wax dummies of the Musée Grevin, and later the cinema: this was a whole new world of the imagination that France contributed to the collective mythology of the modern world.

(Italy too had this kind of raw material: just think of the book by Ernesto Ferrero, *La mala Italia*, which Rizzoli published a few years ago. But we did not have the literary culture – or even just a particular turn of imagination – that was capable of transforming all this.)

However, the 'human interest stories' analysed in the Paris exhibition do not just cover crime stories. Acts of heroism, abnegation, courage, particularly rescues, are also part of the picture. A collection of short works in 1787, on the eve of the Revolution, was dedicated to the 'Virtues of the People': these were human interest stories in which humble people took centre stage, confirming Rousseau's ideas on the natural goodness of human beings.

Not only the extremes of the human mind in terms of good or

evil, but every act that departs from the norm helps to make news, to create a human interest story: the arrival of the first giraffe in Paris in 1827 was an event that for years continued to be illustrated in woodcuts, lithographs, almanacs, on majolica plates, on copper pans.

There are also living phenomena, which from ancient times onwards have carried with them the aura of prodigies, of signs from the gods. The exhibition is not very rich on monsters, mermaids, dwarves, giants or Siamese twins, but there is one exhibit that you certainly don't see every day: a 'naturalized bust' of a bearded woman (of about a century ago), in other words not a representation but the real head of the woman, embalmed after her death for the purposes of scientific documentation. The embalmer, out of both 'artistic' and gentlemanly scruples, has placed round her neck a little collar of embroidered lace.

Also newsworthy are of course incidents and accidents of all types: the rarer and more novel they are, the more they are prized. So we see the first car accidents: a car crashing down on an express train (in America: the backdrop shows rocky mountains, and the vegetation is exotic).

Many of the covers of *Le Petit Journal* show human figures as they fall, suspended in mid-air, or in free fall: a spectator at the theatre plummets from his box down into the stalls, a pilot falls from his balloon, a woman with long skirts flies through a window ('act of madness'), from another window a 'new Icarus' flies, covered in feathers.

And the scenes of violence and crime are always portrayed with raised arms brandishing daggers or knives. The event that shakes the natural order of things is situated in a moment which is as it were outside of time, a fleeting moment that remains fixed for ever.

[1983]

A Novel Inside a Painting

Periodically the Louvre puts on *Dossier* exhibitions, taking a famous painting – or group of paintings – and putting alongside it all the essential documents (drawings, sketches, other works of art) needed to shed light on its genesis. They are always interesting shows, and there is always a lot to learn from them. This winter the *Dossier* exhibition allows us to study one of the most famous paintings of the nineteenth century, figure by figure: Delacroix's *Liberty Leading the People*. A painting with so many people in it is a little bit like a novel where several plots intertwine; that is the reason I feel I am entitled to talk about it, without wanting to invade the field of art historians and critics, but simply recounting what is explained in the exhibition, and trying to read the painting as one reads a book. In July 1830, three days of popular revolt in Paris (the 'Three Glorious Days') had put an end to the rule of Charles X and to the Bourbon restoration; a few days later Louis Philippe d'Orléans's constitutional monarchy was installed; in the final months of the same year Delacroix painted his great canvas in celebration of the July Revolution. Even today, when one needs an image to commemorate the liberating power of a popular revolt, with all the emphasis the theme deserves, people all over the world turn to this painting. The show exemplifies this later reception as well in a room which displays how the painting continues to be referenced, reproduced, caricatured, dressed up in all sorts of different guises: a reception that is certainly due to its theme, but above all to its painterly qualities, which have never been equalled in representations of this type. This work was revolutionary first of all in the history of painting, because, even though nowadays it seems to us an allegorical work, at the time it was seen as the first expression of a 'realism' that was unheard of and scandalous.

Now it has to be said that the painting does not stem from political militancy on Delacroix's part: under Charles X the painter was already at the height of his fame, with supporters at Court and commissions from the State. The scandal caused by his 1827 painting of *The Death of Sardanapalus*, a painting that was judged to be immoral, had ended up by consolidating his reputation. At the same time he also enjoyed the support of the Duke of Orléans, the future Louis Philippe, who was then leader of the liberal opposition and an enthusiast of the new style of painting, and someone who bought paintings from Delacroix.

When the revolt broke out in July 1830, Delacroix did not mount the barricades, but enlisted – like many other artists – for guard duty at the Louvre in order to protect the museum's collections from any danger of being ransacked by the furious mob. We still have some evidence regarding these stints on daytime and night-time guard duty, where furious rows erupted amongst the artists doing their rounds, sometimes ending in fisticuffs, not about politics but about artistic tendencies or how to evaluate Raphael. The image that this scant collection of anecdotes evokes helps us relive that atmosphere of revolutionary tension through an extraordinarily authentic image: the rooms of the Louvre, by night, in the heart of a city in revolt, and these civilians who are armed and cloaked, moving amongst the Egyptian sarcophagi, discussing the ideals of their art with unparalleled fervour, while distant echoes of shots and roars float over from the Hôtel de Ville on the other side of the Seine . . .

Delacroix had interrupted his diary for those years, and to understand his attitude towards the revolution we have some letters where what we see are only the worries of a tranquil man in a time of upheaval. A piece of evidence from Alexandre Dumas (who, however, always distorted his memories) shows us a Delacroix at first visibly shaken at seeing the populace in arms, but then enthusiastic at the sight of the tricolour flying once more as in Napoleon's day, and from that point on he was won over to the people's cause.

In the days following the revolution, the National Guard, which

Charles X had dissolved, was re-established, and Delacroix immediately enlisted as a volunteer, though grumbling in his letters about how tough the turns of duty were. His entire conduct was extremely consistent: his reactions were absolutely normal for someone who sympathized with the way the people's anti-absolutist movement led to the formation of a liberal monarchy, and the logical outcome was for him to end up as part of the new Orléanist establishment.

But 1830 had marked not only the shift from one dynasty to another, and from an aristocracy with bourgeois leanings to a bourgeoisie with aristocratic leanings: for the first time the proletarian masses had taken to the streets *in person* (while even in the 1789 Revolution the initiative had been taken by the ideological leaders), and they had been the decisive element in the change of regime. This *novelty*, which dominated all talk at the time, would be the specific theme of the painting that Delacroix would work on in the months immediately after these events (when already disillusionment and recrimination were seeping through the ranks of the most radical republicans and democrats). 'I've taken up a modern subject, the barricades,' he wrote to his brother in October, 'and if I did not fight for our country, at least I will paint for it. This has put me back in a very good mood.'

A painting that puts before the eyes of the observer the energy, movement and enthusiasm of the event would be thought to have been painted on the spur of the moment. Instead, the history of the work is one of a laborious composition, full of hesitations and about-turns, worked out detail by detail, the artist placing side by side elements that were heterogeneous, some of them motifs from existing paintings. As an allegorical work, one would say that it was inspired solely by an ideal that the painter felt passionately about. Instead, the choice of every detail of clothing, of every weapon brandished, has a meaning and a story behind it. As a realistic work, one would think it was inspired by real life, by emotions aroused by the field of struggle; instead it is a repertoire of quotations from other works in museums, a kind of compendium of figurative culture.

For a start, the exhibition shows, in a way I find convincing, that the female figure at the centre of the painting, the most famous image of Liberty in the history of painting, did not come into Delacroix's imagination only then: she had been present already in a large number of his drawings in the previous ten years or so. She was 'Greece Rebelling against the Turks', a subject to which Delacroix wanted to dedicate a painting from the time of the first stirrings of Greek independence in 1820. The solidarity with Greece was a very powerful theme in European Romanticism, and in 1821 Delacroix began a series of drawings for an allegorical painting. The following year, hearing of the bloody suppression of the revolt on the island of Chios, he developed another idea which would eventually lead to the famous 1824 painting *The Massacre at Chios*. A few years later he would resume these studies for the female figure who would become *Greece on the Ruins of Missolonghi* (1827, now in the Musée des Beaux-Arts in Bordeaux). But more than in these paintings it is in the drawings (which on the basis of other details are recognizable as preparatory studies for *Greece on the Ruins of Missolonghi*) that we recognize the movement of her arms and torso that will be taken up again in *Liberty Leading the People*. Similarly, we see that in these same drawings the figure wears on her head initially a turreted crown, then later a Phrygian cap.

But that is not all. Infra-red photography has been used to determine whether Delacroix by any chance used an already painted canvas for *Liberty*, correcting a draft made years before for *Greece on the Ruins of Missolonghi*. The results of this research are, as often happens, very uncertain, and have led to further questions rather than answers. One thing that is sure is that Liberty's skirt had initially been much fuller, and less suitable for leaping on to the barricade. Her face was initially seen full-on, as in the drawings: the decision to paint her in profile, which gives her that unforgettable incisiveness, seems to have come to the painter as he worked on this canvas, and it is certainly an idea linked to the theme of 1830 and not to the earlier one: Liberty turns her face to the people to exhort them to fight. And then there is a strange cuff on the worker on the left, one of the few points where the painting seems rather botched: this might be a correction to cover a Greek or Turkish outfit. There are also the rickety tables of the barricade, which in the overall composition of the work are used to relegate the landscape of Paris to the background and to raise up as though on a stage the glorified figures in the foreground, but they might also have been used to hide a different landscape that had been painted previously, the beach at Missolonghi, the sea . . .

None of these findings are certain; they remain hypotheses. The only conclusion is that *Liberty Leading the People* is an autonomous painting, devised and painted in 1830; to have used previous studies does not prove any contradiction of, or indifference to, the painting's theme, which is that of the freedom of peoples, but rather enriches it in this move from an ideal allegory to a lived experience which becomes a visual and corporeal fact. The formal composition of the work was determined by the painter in 1830: it is the flag that acts as the apex of the composition, providing it with its triangular structures and the three colours that find a counterpoint in the rest of the painting.

The curators of the exhibition cite as an analogous case that of Picasso, who, on hearing the news of the bombing of Guernica, took up his studies for bull-fights from previous years (which already contained presages of the tragedy that was looming over Spain) in order to compose his famous painting.

Consistently with the theme he had chosen, Delacroix places to the left of Liberty the figures of three workers: what he meant by 'people' were manual workers. (There is not a single recognizable bourgeois person in the painting, except for a figure wearing a cocked hat in the background, who could be one of the students from the Ecole Polytechnique who took part in the uprising.) Driven by a clearly sociological intent, Delacroix portrays three different types of manual workers: the man with the top hat could be an artisan, a *compagnon* from a guild (yes, at that time the top hat was a universal item of headwear, with no social connotations; but the wide trousers and the red flannel belt were typical of the workers); the man with the sword is a worker from the manufacturing sector, with his worker's apron; the man who is wounded, on his hands and knees, with the handkerchief on his head and his blue shirt tucked into his belt, is a manual worker from a building site, one of the seasonal workers who had emigrated from the country to the city.

To the right of Liberty there is the famous urchin with the black beret brandishing two pistols, whom today we all call Gavroche (but in 1830 *Les Misérables* had not yet been written: Hugo's novel would not be published until 1862). In this exhibition he is compared with a statue of Mercury by Giambologna, the élan of which had already inspired other painters of the time, but was always portrayed in profile, whereas here the figure is full-on.

There is another boy armed with a bayonet to the left of the painting, crouching amongst the piled-up paving stones, wearing the beret of the National Guard. All the details of the uniforms are identifiable, so precisely are they represented, and so too are the weapons, from the Royal Guards' bandolier that the urchin has got hold of (though the two pistols were from the cavalry) to the sabre from a crack infantry unit, held by the worker with the apron, along with its own bandolier. You can trace the story of all the weapons in the picture, just like the stories attached to the weapons of the paladins in the chivalric romances; except that here, as always in real revolutions, the weaponry is haphazard and heterogeneous, since it comes from chance seizures and from booty snatched from enemies. The only non-military weapon is the one brandished by the worker in the top hat, which is a hunting rifle.

This detailed iconological analysis has unearthed some ideological surprises regarding this very figure, the most working-class figure here, the man with the apron: on his beret he has a white cockade with a red bow. The white cockade means he is a monarchist, and what is more his pistol tucked into his belt is held by a scarf that has Vendean colours; but he has converted to liberalism (the red bow): so what we have here is a man of the people who is loyal to the throne but is rebelling against the oppression of absolutism ... However, it is pointless immersing ourselves further in these hypotheses where the specialists can tell us whatever they like without fear of us gainsaying them.

More pertinent to our reading of the painting are the three corpses in the foreground. One belongs to allegory and myth, in his idealized, classical pose: in fact he is naked, or at least has lost his trousers, but no critic ever dreamed of being scandalized by this (whereas Liberty's naked breasts, despite being a traditional attribute of Winged Victories, were met with protests) because he follows a widely known academic model from the classical repertoire. That was the pose in which Hector was portrayed when he was tied by his legs to Achilles' chariot. The other two corpses are soldiers of the defeated king, who have been placed along with the fallen of the revolution in a mood of universal *pietas*. One of them wears the uniform of the Swiss Regiment of the Royal Guard (which Louis Philippe would later suppress), and his helmet (the *shako*) has rolled alongside him; the other is a French soldier, a *cuirassier*. These figures of fallen men (including the wounded worker) have precedents in the celebratory paintings of David and Gros: Delacroix's painting is where old and new motifs in the history of painting meet, just as the July Revolution had been a meeting-point in the history of France.

The Parisian landscape in the background raises further problems. It is dominated by the mass of Notre Dame, but the cathedral from that perspective could only be seen from the left bank of the Seine, and in that case one cannot explain the tall houses it has on the right, because on that side it only has the river. In short, what we have here is a fictional, symbolic landscape. Why Notre Dame? The cathedral had nothing to do with Orléanist symbolism (Louis Philippe portrayed himself as a secular person, a son of the Enlightenment) but with the symbolism related to social theories of the time, with the democratic Christianity of De Lamennais; and this was also the period when Victor Hugo began writing *The Hunchback of Notre Dame*, in which the cathedral is a symbol of freedom.

Because of all this, the painting shocked the public and the critics when it was displayed at the Salon of 1831. What aroused protests were the realism of the insurgent proletarians, who were defined by the critics as 'looking like jailbirds', 'scum', 'the dregs of society', and the daring portrayal of Liberty, especially since she displayed one of her armpits as having hair (the classical nude was always hairless). Despite all this, the painting was bought by the Ministry of the

Interior, but already in 1832, when the new regime started to clash with further popular revolts, it disappeared from circulation. It was displayed again after the 1848 revolution, but only for a short time; in 1849 the subject-matter once more became controversial, and the work ended up in a stock-room. The author tried to resurrect it from there for the Universal Exposition of 1855. But Marianne's Phrygian beret was still a subversive image at the time, even though Delacroix had simply followed classical models here, without any partisan intent. Now we can see that the Phrygian beret is not red like those of the *sans-culottes* but of a dark-brown colour. And X-ray scans have discovered that underneath this faded colour there is a layer of scarlet. It is a short step from here to inferring that Delacroix toned down that highly provocative red in order to appease the censors of the Second Empire. Whatever the case, at the request of Napoleon III, the painting ended up in the exhibition. After a number of other vicissitudes, under the Third Republic, the work entered the Louvre, and after that into universal glory.

[1983]

Say It with Knots

In New Caledonia messages of war and peace were transmitted by means of a rudimentary rope made from banyan bark (*Ficus bengalensis*) knotted in various ways. A piece of rope with a bowline knot at one end was a proposal to join a military alliance; if the addressee accepted the alliance, all he had to do was tie a similar knot at the other end of the rope and return it to the sender; in that way an indissoluble pact was agreed. On the other hand, a knot around a small firebrand – extinguished, but with traces of burning on it – was a declaration of war; it meant 'We shall come and burn your huts.' The message offering peace to the vanquished was more complicated; it was a question of persuading them to return to their destroyed village and to rebuild it (the conquerors were very careful not to settle in a village belonging to others and to the spirits of their dead); for that reason the knot for this message would tie together pieces of the reeds, shrubs and leaves that were used for building huts.

These knotted fibres are on display in an unusual exhibition: Knots and Bindings, at the National Foundation for Graphic and Plastic Arts, in Rue Berryer, in Paris. The display invites us to reflect on the language of knots as a primordial form of writing.

These fibres call to mind the pieces of rope used by the Maori (we are still in the Pacific) and mentioned by Victor Segalen in his novel *Les Immémoriaux* (*A Lapse of Memory*): the Polynesian bards or narrators recited their poems by heart, with the aid of interwoven strings, the knots of which were counted between their fingers to mark off the episodes of their narrative. It is not clear what correspondence they established between the succession of names and deeds of heroes and ancestors on the one hand and the knots of different size and shape placed at different intervals along the strings on the other; but certainly the bunch of threads was an indispensable *aide-memoire*, a way of making the text permanent before any form of writing. 'These strands were called "Origin of the Words",' writes Segalen, 'for words seemed to spring forth from them' (*A Lapse of Memory*, translated by Rosemary Arnoux, Brisbane: Boombana Publications, p. 27). The advent of writing, or rather the sole fact of knowing that white men entrusted their memory to black signs on white paper, put the processes of oral memory into crisis: the bards forgot their poems, the strings stayed silent in their hands. 'The oral tradition,' writes Giorgio Agamben, commenting on Segalen, 'maintains its contact with the mythical origin of the word, in other words with what writing has lost and what it continually pursues; literature is the never-ending attempt to recover those lost origins.'

In the exhibition in Rue Berryer there is also a *quipu* from the Peruvian Incas. This is a fringe made up of differently coloured cotton threads used by the high functionaries of the Inca Empire for State accounts, population censuses and evaluation of agricultural produce: in short, it was the computer of that society, and everything was based on the precision of its calculations and measurements.

There is a Japanese object made from thin strips of wood knotted in a complicated, almost baroque pattern, which symbolizes the god of the mountain who flees to the mountain-tops during winter in order to descend on to the plain in spring as the god of rice and to watch over the young plants. In the tradition of Japanese Shintoism there are gods called 'knotters' because they bind the heavens to the earth, the spirit to matter, life to the body. In their temples a cord of knotted straw indicates the purified space, closed off to the profane, where the gods can stay. In the more sophisticated Buddhist rituals, the power of the knot exists even without any material support: all the priest has to do is to move his fingers as if tying a knot for the space of the ceremony to be closed to malign influences.

There is only a small number of such ethnographic exhibits, lent by the Musée de l'Homme, the Museum of African and Oceanic Arts and private collections, and there are also items from the Museum of Popular Arts and Traditions. In fact the main part of the exhibition consists of knots in the artworks of contemporary artists; their work contains loops, knots and tangles in a large variety of materials inspired both by the primitive force of anthropological objects and a creative fascination in the countless practical uses of knots in everyday life.

Without wishing to invade the field of art criticism, I will briefly mention the following items: a beautiful ensemble by Etienne-Martin (ropes, belts, horse caparisons, mats); a barrier made of poles, ropes and rolled-up tents by Titus-Carmel; a palisade held together by tangles of hemp by Jackie Windsor; a gravel bed with remains of carbonized ropes resting on it by Christian Jaccard; many bewitching coloured objects by Jean Clareboudt; bows decorated with ribbons by Louis Chacallis; bindings round lead pipes by Claude Faivre; roots made from hawsers by Danièle Perrone; and other examples of knotty materials from nature (a root, a bird skeleton by Louis Pons, entangled vegetable fibres by Marinette Cucco).

One of the cases in the exhibition induced in me, as a professional writer, a rather nightmarish sensation: 'Imprisoned Books' consisted of volumes that were tied up, gagged or hanged, in all sorts of ways, a book wrapped up in a mass of hemp and lacquered in the colour of a red lobster (Barton Lidicé Beneš), or – a less oppressive spectacle – a book with pages made of gauze like embroidered spiders' webs (Milvia Maglione).

The exhibition, organized by Gilbert Lascault, has a catalogue in which there is an article by a mathematician, Pierre Rosenstiehl. The fact is that knots, as linear configurations in three dimensions, are the object of a mathematical theory. Among the problems the theory opens up are those relating to the Borromean Knot or Chain (three rings linked together, but any two of the rings are linked only by the third). The Borromean Knot was also very important for Jacques Lacan: see the chapter 'Ronds de ficelle' (Knots of Rope) in his *Séminaire, Livre XX: Encore, 1972–1973* (Paris: Seuil, 1975), pp. 107–23.

I would never dare to try defining in my own words the relationship of the Borromean Knot to the unconscious according to Lacan;

but I would like to try to formulate the geometrical and spatial idea I have managed to draw from it: the three-dimensional space actually has six dimensions since everything changes according to whether one dimension passes above or below another, or to the right or left of the other, just like in a knot.

This is because in knots the intersection between two curves is never an abstract point but is the actual point where one end of a rope or cord or line or thread or string either runs or turns or is tied above or below or around itself or around another similar item, as a consequence of very precise actions carried out by practitioners of a range of crafts, from the sailor to the surgeon, the cobbler to the acrobat, the mountaineer to the seamstress, the fisherman to the packer, the butcher to the basket-maker, the carpet-maker to the piano-tuner, the camper to the chair-mender, the woodcutter to the lace-maker, the bookbinder to the racquet-maker, the executioner to the necklace-maker . . . The art of making knots, which is the peak of both mental abstraction and manual work, could be seen as the human characteristic par excellence, just as much and perhaps even more than language . . .

[1983]

Writers Who Draw

With the arrival of Romanticism, writers in France began to draw. The pen runs across the page, stops, hesitates, absent-mindedly or in great agitation leaves a profile in the margin, or a puppet, or a scribble, or it becomes engrossed in drawing a frieze, a piece of shading, a geometric labyrinth. The thrust of graphic energy finds itself every so often faced with a dilemma: to continue evoking its own ghosts through the steady trickle of the uniform letters of the alphabet or to pursue them in the visual immediacy of a rapid sketch? It appears that this temptation did not always arise: there have always been painters who write, but very few writers who draw. All of a sudden, between the end of the eighteenth and beginning of the nineteenth centuries, the education of a young person who would become a man of letters was not considered complete if it did not include some training in drawing or painting. The lives of poets and writers opened up to a practice and ambition which in some cases could have led to a professional commitment in the field of art, if the other vocation had not been stronger. At the same time even the manuscripts of those who had no artistic education began to swarm with small figures or doodles. It was the cultural physiognomy of the writer that was changing, with the new aspiration towards the 'total work of art', which took shape in Germany under Romanticism, a dream cherished by Novalis (who invented the phrase), and which would be the basis of Wagner's programme. Hoffmann (who was translated into French in 1829) instantly became a model for new French literature, not only because he created a new genre, the *Contes fantastiques* (it was the French, always ready to provide a label for cultural innovations, who invented the term, which had no equivalent in German), but also because he was presented as someone who was both a writer, a draughtsman and a musician: the new type of versatile talent which Romanticism had called forth.

These ideas were stimulated by an exhibition in the Maison de Balzac, Drawings by French Nineteenth-Century Writers. It displays 250 documents (from simple doodles to sketches and caricatures or indeed to watercolours and actual paintings) by forty-five famous or minor or forgotten poets and writers, but all of them significant for the relationship between pictorial penmanship and writing. It has to be said at the outset that the theme is valid only in the most general terms, for the claim that one can establish a link between the style of a given writer and their drawings seems untenable given the very evident absence of style in the drawings themselves, whether this is due to the author's crude hand or to a skill that is too impersonal. So I believe that it is impossible to establish why many writers draw and many others, despite their works being full of visual imagery, do not draw at all. (The list of non-drawers is impressive too, including as it does Chateaubriand, Madame de Staël, Flaubert, Zola.)

It has been known for some time that the most talented amateur painter amongst French nineteenth-century writers was Victor Hugo, and this exhibition confirms it. For the show they have transported from the Maison Victor Hugo (the other – and more interesting – writer's house that is now a museum in Paris) some of his haunting works in pen and ink depicting ghostly cities and eerie landscapes, where the artist in those anxious years gave vent to the darkest vein in his Romanticism, but also displayed an ingenious inventiveness in his experimentation in this medium.

As for the other champion, also in terms of written output, Balzac, he had no talent at all for drawing, and limited himself to scribbling a few small, rather childish drawings (especially faces) in the blank spaces in his manuscripts. Balzac, then, despite the fact that the exhibition is in his house, is represented by only two pages, and they are reproductions at that, not even originals. Another lacuna is Stendhal, but knowing the rudimentary sketches that accompany his *La Vie de Henry Brulard*, we could almost consider him as belonging to the group of non-drawing writers. Neither was

Michelet very skilful with a pencil, judging by a very summary sketch for a proposal of his for a monument to the fallen of the French Revolution.

Then there are the writers who can draw all too well and who for that reason are less interesting. Mérimée, Alfred de Vigny, Théophile Gautier had had proper art lessons, but the many works on show here – illustrations of historical subjects, watercolours of landscapes, caricatures, architectural drawings – seem rather anonymous. Even the drawings with which Mérimée filled the headed ministerial notepaper during the meetings of the many official commissions on which he was an important member are very composed and academic. More interesting are his travel notebooks, for their precise observation of countries and costumes, things which we find forcefully expressed in a very different way in his short stories. Amongst those by Gautier two drawings in red ink stand out, testimonies to his grotesque, tormented-poet tastes: a witch's kitchen and a Temptation of St Anthony in an erotic and sadistic vein.

George Sand was a technically accomplished landscape artist in pencil and watercolours, but at least in a group of grey-green, light-brown mountain-views she manages to convey something unusual: a bristling, stagnant, mineral wasteland. These are little scenes created using a technique she herself invented; she called them 'dendrites', after those stones which exhibit a faint pattern of branching, multicoloured veins.

The most unexpected discovery of the exhibition is Alfred de Musset as a precursor of comic strips. This 'child of the century' of Romanticism, to quote his autobiographical work, *La Confession d'un enfant du siècle*, would compose for his own entertainment, as well as for that of friends and family members, stories in the form of cartoons with caricatures of well-known characters: on display here are two complete sets of stories. One tells the tale of a journey to Sicily by the poet's brother, culminating in an affair with a lady of easy virtue from Messina. The other is the account of a piece of Parisian gossip: how a large-nosed gentleman proposed to the singer Pauline García (sister of Maria Malibran) and how during the successive bouts of breaking-up and making-up between the couple the fiancé's conspicuous nose changed shape and size. The great thing is that another contender for the singer's hand was the poet himself, Alfred de Musset, who portrays himself as a patient confined to bed by a pulmonary illness, undergoing improvements and relapses in his condition which are caused by the alternating fortunes of his rival. Even in caricature Pauline is not lacking in a certain dreamy grace, but the evil brains behind the plot was George Sand, portrayed with her cigar or a long pipe and brandishing a sabre.

These comics before their time are of a surprising modernity in their narrative conception and graphic elegance, something between Rodolphe Töpffer and Edward Lear, so much so that they attain a stylized liveliness that looks almost twentieth-century, reminiscent of Sergio Tofano's drawings. With de Musset we see the start of the custom of illustrating letters to a lady friend with little drawings (theatre gossip once more, where the same characters keep resurfacing). De Musset is one of the cases where one can talk of 'writer's drawings' as of something different from an artist's drawings, in that they are connected with a narrative creativity and stylization and also have a kind of irony and self-mockery: all of these are literary procedures, even though they are noticeably different from the procedures used by the author in his written works.

The other type of 'author's drawing' that the exhibition highlights is that of writing that becomes drawing, and here the surprising example is Barbey d'Aurevilly, who kept an illustrated diary or scrapbook using different-coloured inks, where the written phrases are interspersed with arrows, hearts, suns, chalices, geometric decorations, all of them rather rudimentary and disorganized but with a tremendous graphic vitality and brio bordering on the effects of 'Outsider Art'. This great French dandy possessed a complete cache of coloured inks and goose-quill pens of varying sharpness as well as paintbrushes. For instance, he would rewrite in gouache his stylized signature, which he had already written in pen, until it became a dense and viscous calligram, or he would compose abstract hieroglyphics that resembled monstrous insects or mobiles in the air.

Baudelaire knew not only how to draw but how to put his intelligence at the service of his pencil (or his charcoal or his ink), and his self-caricatures have a pungent sharpness. The epoch that starts with him, in other words the second half of the century, sees poets and writers becoming more relaxed, less academic, in drawing figures on paper (apart from those who were painters as their first profession, such as Fromentin, or those who did etchings following all the rules, such as Jules de Goncourt, or those who illustrated their exotic voyages with accurate devotion, such as Pierre Loti).

More than the novelists (Dumas the Younger was a good caricaturist, Maupassant drew droll puppets, Anatole France was a draughtsman of elegance and bravura), it is the poets who catch our attention, especially Verlaine, who had never studied drawing but who was a witty draughtsman, full of fun and inventiveness and with a very modern touch. There are many self-portraits that caricature the jutting-out jaw beneath his minuscule nose, emphasizing that Mandarin Chinese look he had: that is how he appears in a small sheet of paper where his features are simplified into a series of superimposed triangles before being totally deconstructed into pre-Cubist planes. But the most moving piece is a portrait of Rimbaud leaning across a café table staring at a bottle of absinthe, with the face of a child in a sulk. (But Rimbaud himself was not an interesting draughtsman, at least judging by the two examples exhibited here.)

A poet who put special calligraphic care into his letters was François Coppée, who structured the page of every letter with great clarity and illustrated it with ideograms and rebuses. In his love letters to Méry Laurent (a demi-mondaine who was maintained by an American dentist) he called his girlfriend 'dickie-bird' and himself 'the she-cat'. However, these nicknames, which to our ear sound as if they have got the genders wrong, are systematically replaced by the drawing of a dove preening itself and a puffed-up tom-cat, who imprint themselves on our eyes as female and masculine ideograms respectively through the evocative power of the drawing.

Méry Laurent was courted in the same period by Mallarmé as well, who also wrote her letters with little drawings. He too portrayed her in ornithological terms, but used far more ink: for Mallarmé she was 'the peacock'. Mallarmé was not at all good at drawing, nor had he developed any technique, but he put into his figures something of the great delight that animated his incomparable verbal talent. One note in which he makes an appointment with 'the peacock', who was arriving by train, becomes a precious Mallarmean 'comic strip', scribbled with great joy.

The pursuit of a sphere of expression different from that of words is the urge behind many of these pictograms which have been traced in the margins of pages that are dense with writing. How can one not feel the eternal, irrepressible envy of the writer for the painter? 'What a happy job is that of the painter compared with that of the man of letters!' we read in the Goncourt brothers' journal, dated 1 May 1869. 'In the one we have the happy activity of the hand and eye as against the torment of the brain in the other; and the work that for one of them is a joy is torture for the other . . .'

[1984]

II The Eye's Ray

In Memory of Roland Barthes

Amongst the first details to emerge about the road accident on 25 February at the junction of Rue des Ecoles and Rue Saint-Jacques was that Roland Barthes had been disfigured, so much so that nobody, just a few yards from the Collège de France, had been able to recognize him. The ambulance that picked him up had taken him to the Salpêtrière Hospital as if he were an anonymous victim (he had no documents on him) and so he stayed in the ward unidentified for hours.

In his last book, which I had read just a few weeks previously – *La Chambre claire, Note sur la photographie* (Editions Cahiers du Cinéma-Gallimard-Seuil, 1980) (*Camera Lucida. Reflections on Photography*, trans. Richard Howard, London, Vintage, 1993) – I had been struck above all by the wonderful pages on the experience of being photographed, on the unease caused by seeing one's own face become an object, on the relationship between the image and the ego. As a result, amongst the first thoughts I had as I waited anxiously to hear his fate was the memory of that recent reading, the fragile and distressing link with his own image which was suddenly being torn apart just as one tears up a photograph.

However, on 28 March, in his coffin, Barthes's face did not seem at all disfigured: it was him, exactly as he was when I regularly encountered him in the streets of the quartier, cigarette dangling from a corner of his mouth – that was the way all those who grew up before the Second World War smoked (the historicity of the image, one of the many themes of *Camera Lucida*, applies also to the image that each one of us gives of ourselves in life). But his face was fixed like that for ever, and the same pages from chapter 5 of the book, which I went back to reread immediately, now spoke of that, only of that, of how the fixity of the image is death, hence our inner reluctance to be photographed, as well as our submission to it. 'As if the (terrified) Photographer must exert himself to the utmost to keep the Photograph from becoming Death. But I – already an object, I do not struggle' (*Camera Lucida*, p. 14). An attitude that now seemed to find an echo in what we managed to find out about his condition during the month he spent at the Salpêtrière unable to speak.

(The mortal danger he was in was apparent immediately, not in the fractures to the head but in those to his ribs. So then another quotation immediately occurred to his distressed friends: the one about the rib that had been amputated in his youth because of a pneumothorax and which he kept in a drawer, until he decided to throw it away, in *Roland Barthes by Roland Barthes*, trans. Richard Howard, New York: Hill and Wang, 1977.)

These references in my memory are not accidental: the fact is that his entire oeuvre, I now realize, consists in forcing the impersonality of our linguistic and cognitive mechanisms to take account of the physicality of the living and mortal subject. Critical discussion of Barthes – which has begun already – will be polarized between the supporters of one Barthes or the other: the one who subordinated everything to a rigorous methodology and the one who had just one sure criterion, namely pleasure (the pleasure of intelligence and the intelligence of pleasure). The truth is that those two Barthes are just one: and in the constant and variously balanced co-presence of these two aspects in Barthes lies the secret behind the fascination his mind exercised on many of us.

On that grey morning I was wandering through the empty streets behind the hospital, looking for the 'lecture theatre' where I had learned Barthes's body would leave from, in a very private ceremony, to go to the provincial cemetery where he would join his mother in her grave. And I met Greimas, who had also arrived early, and who told me about how in 1948 he had met Barthes in Alexandria, in Egypt, and had made him read Saussure and rewrite *Michelet*. For Greimas, an unbending master of methodological rigour, there was no doubt: the real Barthes was the one who carried out the semiological analyses with discipline and rigour as in *Le Système de la Mode* (*The Fashion System*); but the real point that made Greimas disagree with the obituaries in the newspapers was their attempt to use professional categories like philosopher or writer to define a man who eluded all classifications because everything he had done in his life he had done for love.

The day before, when François Wahl told me the time and place of that almost secret ceremony, he had spoken of a 'cercle amoureux' of young men and girls which had formed together round Barthes's death, a circle jealously possessive of a grief that would not tolerate any other display but that of silence. The stunned group which I joined was made up to a large extent of young people (few of those in their midst were famous; but I did recognize Foucault's bald cranium). The plaque on that wing did not have the university denomination 'Lecture Theatre', but said 'Salle de reconnaissances' (Mortuary Chamber), and I realized that it must indeed have been the morgue. From behind white curtains that went all around the room every so often a coffin would come out, shouldered by pall-bearers all the way to the hearse and followed by a family group of ordinary people, tiny little old women mostly, each family identical to the one in the preceding funeral, as if in a superfluous demonstration of the levelling power of death. For those of us who were there for Barthes, waiting motionless and silent in the courtyard, as if following the implicit message to keep the signs of a funeral ceremony to the minimum, everything that appeared in that courtyard seemed to magnify its function as sign; I felt that over every detail of that wretched scene hovered the sharp gaze that had trained itself to discern revelatory clues in photographs in *Camera Lucida*.

So this book of his, now that I'm rereading it, seems to me to point entirely towards that final journey, that courtyard, that grey morning. Certainly it was from an examination of the photographs of his recently dead mother that Barthes's reflections in the book had started (as is recounted at length in the second half of the volume). It was an impossible pursuit of his mother's presence, which he found in the end in a photograph of her as a little girl, an image that was 'a lost, remote photograph, one which does not look "like" her, the photograph of a child I never knew' (p. 103). The photograph was not reproduced in the book, because we could never have understood the value it had taken on for him.

A book about death, then, just as the previous one – *Fragments d'un discours amoureux* (*A Lover's Discourse: Fragments*) – had been a book about love. Yes, but this one too was a book about love, as is proved by this passage about the difficulty of avoiding the 'heaviness' of one's own image, the 'meaning' to give to one's own face: 'For it is not indifference which erases the weight of the image – the Photomat always turns you into a criminal type, wanted by the police – but love, extreme love' (p. 12).

This was not the first time that Barthes had spoken about being photographed: in his book on Japan – *L'Empire des signes* (*Empire of Signs*) – one of his least-known works, but one of the richest in terms of acute observations, there is the extraordinary discovery, when observing photographs of himself published by Japanese newspapers, of a look in his features that was indefinably Japanese. This was explained by the way they have over there of touching up photographs, making the pupil of the eye round and black. This discourse about intentionality being superimposed on our image – historicity, belonging to a particular culture, as I said before, but above all the intentionality of someone who is not us but who uses our image as a weapon – crops up again in *Camera Lucida* in a passage on the power of *truquages subtils* ('the subtlest deceptions', p. 14) used on photographs. Barthes had discovered a photograph of himself, one where he thought his grief for a recent death was recognizable, on the cover of a book that satirized him where his face had now become detached from his inner self and looked sinister.

The reading of this book and the death of its author happened too close to each other for me to be able to separate them. And yet I must succeed in doing so in order to give an idea of what the book really is: it consists of Barthes's progressive approach towards a definition of that particular type of knowledge that was opened up by the advent of that 'anthropologically new object' (p. 88), the photograph.

The reproductions in the book are chosen on the basis of this argument, which we might call a 'phenomenological' one. In the interest that a photograph arouses in us Barthes distinguishes between a level that is that of the *studium* or cultural participation in the information or emotion that the photograph conveys and that of the *punctum*, or in other words the surprising, involuntary transfixing element that certain images communicate. Certain images, or rather certain details of images: the reading that Barthes offers of the works of famous or anonymous photographers is always startling. The things he reveals to us as being extraordinary are often physical details (hands, nails) or details of clothing.

Against recent theorizations of photography as a cultural convention, artifice, non-reality, Barthes privileges the 'chemical' basis of the operation: a photograph is the trace of luminous rays emanating from something that exists, something that is there. (And this is the basic difference between photography and language, which can talk of something that is not there.) In the photograph we are looking at something that has been and is no longer there: this is what Barthes calls the *temps écrasé* ('defeated time', p. 97).

It is a typical Barthes book, with its more speculative moments where it seems that by dint of multiplying the links in his terminological net he can no longer succeed in extricating himself, but also with sudden illuminations, flashes of proof that arrive like surprising but definitive presents. From its very first pages *Camera Lucida* contains a declaration of his ongoing methodology and programme of work: refusing to define a 'photographic universal' (p. 5), he decides to take into consideration only the photographs which he was 'sure existed for me' (p. 8).

'In this (after all) conventional debate between science and subjectivity, I had arrived at this curious notion: why mightn't there be, somehow, a new science for each object? A *mathesis singularis* (and no longer *universalis*)?' (p. 8).

Roland Barthes cultivated the science of the uniqueness of every object, in a way that combined a scientist's ability to produce general rules with a poet's attention for what is singular and unique. This kind of aesthetic philosophy or delight in understanding is the great thing that he has – I would not say taught us, because it is not something that can be taught or learned – that he has proved is possible: or that it is possible to search for it.

[1980]

Day-flies in the Fortress

A swarm of day-flies flew into a fortress, came to rest on the ramparts, attacked the keep, and invaded the sentinels' patrol path and the dungeons. The network of nerves on their transparent wings hovered amidst the stone walls.

'It is pointless for you to stretch your wiry limbs,' said the fortress. 'Only those who have been made to last can claim to exist. I last, therefore I am. You don't.'

'We inhabit the space of the air, we mark time with the beating of our wings. What else does existence mean if not this?' answered those fragile creatures. 'You, on the other hand, are only a shape planted there to mark the limits of space and time in which we exist.'

'Time flows over me, but I remain,' insisted the fortress. 'You only graze the surface of coming into being just as you graze the surface of water in streams.'

And the day-flies: 'We dart about in the void just like writing on a blank page and the notes of the flute amid the silence. Without us there is nothing but the all-powerful and ever-present void, so heavy that it crushes the world, the void whose annihilating power is clothed in solid fortresses, the void-fullness that can only be dissolved by what is light and swift and slender.'

You can imagine this dialogue taking place in the Forte del Belvedere in Florence, which is host to Fausto Melotti's aery sculptures, one of which has this very title, *Gli effimeri* (*Ephemera*): it looks like a musical score of ideograms as weightless as water-insects that seem to whirl around a brass bedstead screened by a mesh of gauze.

You can also imagine, if you wish, that the background to this dialogue is the debate on the aesthetics of the ephemeral that we have been hearing about this year in Italy. But you can also easily ignore those discussions, for Fausto Melotti's discourse is about something else: his use of poor and perishable materials – little sticks of welded brass, gauze, little chains, foil, cardboard, string, iron wire, chalk, rags – is the fastest way to reach a visionary realm of marvels and splendours, as children and Shakespearean actors well know.

On the other hand, one cannot help remembering that this exhibition unfortunately is already closing on 8 June, as if by some literal misinterpretation of the value of the ephemeral by the Florentine organizers. As a result, all these works, many of which have been completed specifically for the spaces of the Fortezza del Belvedere (whether they are new creations or 'enlargements' of previous works), will be on display only for two months. This temporally contracted celebration is yet one more paradox in the story of an artist who only in his declining years has been recognized as one of the greats.

On the green bastions of the Belvedere a palisade of geometric shapes made from burnished steel and bristling with spears whose tips are raised in the air or stuck into the ground might suggest a barbaric or extra-terrestrial war; but we instantly realize that it has been placed there to defend a space in which the force that wins out is inner strength, the resistance is one composed of slender lines, the confrontation is backed up by irony.

I would say that it is in dimensions around a metre in height that the rhythms of Melotti's imagination and their placements in the Fortress come together most happily. And happiness here means the maximum of cheerfulness and melancholy together, as in *The Traveller*, who with his string scarf is walking past a wall full of metallic, geometric posters under a sky of textiles.

Or as in *Ulysses' Ship*, which is hollowed out like the breast-bone of a bird, with a little chalk head on a mast. (It must be pointed out that Melotti, one of the 'founding fathers' of abstract art, almost always puts some figurative element, no matter how small, into his works, as though implying that 'rigour' is never where we most expect to find it.) Or like *The Grand Canal*, which is made of perforated bricks placed on a mirror.

'There is love and there is respect for the material,' writes Melotti in a little book of aphorisms entitled *Linee* (*Lines*), published by Adelphi in its Piccola Biblioteca Adelphi series. 'Love is a passion, it can turn into hate: this is a revitalizing drama for an artisan-artist. Respect is like a legal separation: the material demands its rights and everything finishes in a frosty relationship. The true artist does not love or respect his material: it is always "on trial" and everything can go completely wrong (Leonardo, Michelangelo and his works in marble).'

[1981]

The Pig and the Archaeologist

The big news this year, in the excavations at the Roman villa at Sette-finestre near Orbetello, is the pig-sty. This is a courtyard which on each of its four sides has many separate compartments separated by little walls and with hollowed-out areas on the ground for the troughs: these were covered by a portico, the only remains of which now are the bases for the supports. As soon as this structure came to light, the first idea was that in each compartment a pig was kept for fattening; and a pig-farmer, asked about this, recognized it as a structure not unlike those used nowadays. But a reading of the classical sources instantly scotched this hypothesis.

Columella's treatise on agriculture, which belongs to the same period as the villa (first century BC), has a chapter on the rearing of pigs but it never mentions pigs being fattened: it lists the best foods for pigs, but it always refers to their eating while in pasture in the woods. Instead the location of this pig-sty made it clear that it was for sows during pregnancy, as well as for the birth and suckling of the young.

'Pigs should not to be housed together, in the manner of other herd animals,' writes Columella. 'Their sties should be built in the fashion of colonnades, in which the sows can be contained after far-rowing and even while pregnant; for sows, more than any other animals, when they are packed together tightly in a disorderly fashion, can lie on top of each other, thus causing abortions. Therefore, as I have said, sties should be constructed with party walls between them, each four feet in height, so that the sow cannot jump over them. They should not have roofs, so that the swineherd can look in from above to count the number of piglets or to check whether any sow is lying on top of her litter and crushing one of them, in which case he can pull it from under her' (Columella, *On Agriculture*, 7.9.9–10).

The excavations at Settefinestre have thus brought to light a pig-sty that corresponds exactly to the description in Columella, that is to say a huge labour ward for the production of piglets, each sow having a pen (in Latin *hara-harae*) to herself. In fact a fundamental difference distinguishes modern pig-rearing, which aims primarily at fattened pigs, from its Roman equivalent, which was geared primarily towards high numbers of pigs and their ability to move around. The reason for this was that the pigs were not slaughtered at the villa: they had to reach the city on their own legs, in huge groups (just like cattle in the Far West, which were accompanied by cowboys all the way to the abattoirs in Chicago, before the invention of freezer wagons). Thus while the male pigs always lived and fed in the open, the pens or *harae* were reserved for the sows during the four months of their gestation period and the three weeks of suckling. In the pig-sty at Settefinestre there are twenty-seven such *harae*: counting twenty-seven sows capable of bearing eight piglets each time they are brought to birth, and the fact that they give birth twice a year, one can work out that the production rate here was around 400 head of pigs a year.

However, suckling the young presented problems: not only for the Roman pig-rearers but also for today's archaeologists. Columella recommends that each sow should suckle only her own young, because when the piglets get mixed up, they start sucking the teats of any sow, and since the mothers also make no distinction between their own young and those of others, some sows would have been absolutely exhausted, and the more greedy piglets would have been hyper-nourished while others would be dying from lack of food. So the most precious skill the guardian possessed, according to Columella, was memory: the ability to recognize the young of each sow and avoid confusion. An extremely difficult task: one can be helped by placing a sign made with pitch on each suckling pig from the same brood, but 'the most convenient solution is to equip the sties (that is, the *harae*, or the individual pens) with thresholds that are low enough for the sow to step over but too high for the suckling pig to climb' (Columella, *On Agriculture*, 7.9.13).

Here Columella no longer agrees with the site at Settefinestre, where they discovered low thresholds; and he does not agree even with Varro (whose treatise *De re rustica* is no less detailed and precise than Columella's), for Varro says that the thresholds have to be low, otherwise the pregnant sows would crash into them with their bellies and abort. (But Varro did not agree even with himself since a few lines later he too mentions a high threshold to prevent the piglets escaping.)

There is only one way to solve all these contradictions: to excavate and carefully bring to light even the smallest details. In fact it turns out that the thresholds of these *harae* have a groove running along them that is not found in the stone of any other threshold here. What else could this rut be for except for inserting a barrier made of vertical planks, a sliding door that the guardian could lift up to let the sow through but then lower in order to keep the piglets from escaping. Thus the thresholds were low or high according to needs. Thus also, by wielding the trowel with respect for every trace of life as it is lived, archaeological tests show that the facts are not in contradiction with the classics, and not only this but also that the classics are not in contradiction with themselves.

Beneath the ground nothing is lost, or at least the maximum amount of information is preserved; but it is in the act of digging that one risks destroying what the centuries have preserved for us, if the correct technique is not used. Italian archaeology has always had a tendency to privilege the architectural and monumental: it is inspired only by triumphal arches, columns, theatres, baths, and it considers all the rest as unimportant fragments. In countries which are less rich in monumental remains a different school has emerged which has now spread throughout the world and which in Italy has a passionate apostle in Andrea Carandini: archaeology as the search in every layer of the earth for minimal signs and clues from which we can reconstruct daily, practical life, business, agriculture, the phases of the history of society. This is an approach that consists entirely of hypotheses and tests, which proceeds by trial and error, by enigmas and deductions and inductions, as in the case of the pig-sty.

Every summer for the last five years about fifty young Italian and British students have been helping with the excavations at Sette-finestre under the direction of Andrea Carandini. They are male and female archaeology or conservation students who are volunteers completing a placement. Every morning you see them working with picks and shovels and brushing shards of pottery under the sun for eight hours (on site the working day goes from six in the morning until half past two in the afternoon), working with the kind of enthusiasm in the face of hard labour that one usually finds only in activities that offer immediate gratification. To be precise I have to say that the first thing that strikes you is that it is the girls who are hammering with picks and spades and pushing laden wheelbarrows, while it seems as if the boys prefer calmer, lighter work. However that may be, seeing them all together, one has an impression of young people today that is different from the one usually offered by newspapers, but which perhaps best represents the many things towards which people aspire today: collective effort and individual accomplishment, concentration and noncha-lance, alacrity and relaxation.

The secret weapon or symbolic emblem of the new archaeology is a trowel that is much smaller than the one used by Italian builders but which is currently favoured by British ones. The technique of digging without causing disasters, favoured by British archaeologists, perhaps stems from the fact that they had this simple tool to hand. Not having an Italian verb as handy as this tool, at Settefinestre they have coined a verb 'trowelare', from the English word *trowel*.

The fragments that are the result of the collapses that have occurred over the centuries are brought to light layer by layer, then drawn and photographed just as they have been found, described on meticulous index-cards, then taken away and placed on plastic trays in the arrangement in which they were discovered. They might be bits of tiles from a collapsed roof, pieces of frescoed plaster from the walls or ceiling, broken crockery, all the way down to mosaics from floors. Then in a laboratory (at the University of Siena) they are classified and numbered, and they start piecing the puzzle together again.

The villa at Settefinestre has very many things to tell us about Roman society and economy in the republican and imperial eras, and not all of them were even hidden underground. What has remained visible for twenty-one centuries must have been the first thing that one saw even then, travelling along the Roman road at the bottom of the valley: a surrounding wall crowned with towers (fake towers to provide an almost theatrical illusion that these were the walls of a city in the distance). This wall surrounded a garden at the end of which rose up a monumental façade, with a portico and above it a loggia with a panoramic outlook (the loggia had collapsed and the columns had been removed, but the arches of the portico were still visible: that was why this mound was given the name Settefinestre, 'seven windows').

The monumental aspect of the villa, in a position that dominates those areas that were initially and would later still be amongst the most desolate in the Maremma, was meant to stress the importance of the family that had invested capital and slaves in a huge business producing and exporting wine and olive oil. We are near the town of Cosa (the Roman town later identified with the supposed town of Ansedonia), in whose harbour have been found huge quantities of wine amphorae with the mark of the Sestius family. The same mark can be found on amphorae found in the remains of Roman ships on the French and Spanish coasts, as well as in archaeological excavations in the Rhône and Loire valleys. Since the same initials are found on objects from Settefinestre, this seems to prove that the villa belonged to the Roman senatorial family of the Sestii. Partisans of Sulla, the Sestii benefited from the confiscation of land after the civil war between Marius and Sulla, and established themselves in the area round Cosa, where a first phase of agriculture by military settlers (this was the 'centuriation' of lands distributed to the soldiers) had for some time been in a state of crisis and then stagnation. The great technological revolution of intensive agriculture was made possible by the availability of slaves who were prisoners of war and belonged to a few wealthy families, thus allowing the training of a specialized workforce.

We know very little about the life of slaves from written

accounts, and one of the main sources of interest in Settefinestre lies in the information we can deduce from that part of the villa that was reserved for slaves, a self-standing wing that was however joined to the core of the building and divided into cells: a layout not dissimilar to that found in soldiers' quarters, in those sites where there are traces of Roman military camps. It is calculated that each cell could have housed four slaves, and, judging from the part of the villa that has been excavated so far, it can be said that it had about forty slaves, a figure that matches evidence from authors of the time. For the time being we cannot say whether their life was that of prisoners who were housed as in barracks without women, or whether each of these cells could hold a family with a wife and children (thus constituting a kind of slave-farm, given the huge convenience of being able to multiply this fundamental source of energy). The excavations in the slave quarters have less to 'say' than those in the masters' quarters, because the walls have no frescoes or ornaments and there are very few fragments of objects. A ceramic cup with the name *Encolpius* in graffiti on it is one of the few messages the slaves have sent us.

Thus one building contained a luxury residence for the masters, barracks for the slaves and a working farm business (the cellars with the wine and oil presses have been excavated entirely and have allowed us to understand the techniques used). The Roman 'villa' was a productive unit: each villa was in charge of about 500 acres of arable land (roughly 125 hectares). The Sestii certainly possessed several villas in the area, but this one, it seems, had a special role as a residence and as a showcase for their wealth. The management of the land made the presence of the owners necessary for at least part of the year; for that reason the residential area had to allow them a high quality of life, one that did not make the owners long for the comforts of the city.

That is why we find the panoramic loggia which links to an internal colonnaded garden through a room called the *exedra*; an atrium with an *impluvium* to collect rain-water and a mosaic floor; three or perhaps four *triclinia* or dining-rooms with frescoed walls, each one used for one season of the year; six living rooms, amongst which is a rare example of a 'Corinthian room'; four bedrooms, each with two alcoves and bases for wardrobes. As for the garden, we can work out its shape and the position of the flower-beds, since the latter were dug out from the rock and then filled with humus.

On one slope of the hill an area of about one hectare is surrounded by a high stone wall: it is probably a *leporarium*, that is to say a reserve for wild animals such as hares, boars, roe-deer, guarded by slaves who were also hunters. According to Varro, this kind of game-rearing was one of the luxuries of the time. It also acted as a place for displays: Varro tells of a landowner who exhibited himself on his land dressed as Orpheus, surrounded by deer and bucks.

The period in which the villa functioned, according to Andrea Carandini, lasted little more than two centuries, from the first half of the first century BC to the beginning of the second century AD; but the period of its greatest splendour was certainly shorter than that. One could say that Italy's economic and not only economic decadence began at the zenith of the Roman Empire, when it was the most distant imperial provinces and not the Italian peninsula that created the riches that Rome absorbed. Archaeological evidence shows that at Settefinestre already in the first century AD the agricultural installations extended to parts of the villa that had once been residential, a sure sign that the owners no longer lived there. The production of wine and oil was now geared to local consumption and not for export any more. In the first century AD the large landed estate typical of the Empire incorporated the villas of the aristocracy: the cultivation of cereals and pasture land, which both require a less numerous and less specialized workforce, replaced vines and olive groves.

Bit by bit the vaulted roofs and frescoed walls collapsed, the wine-presses were dismantled, the cellars became deposits for grain. The villa was abandoned and stripped of its goods; families of shepherds took refuge in it. Two skeletons from the early Middle Ages buried beneath the portico provide evidence of a humanity now characterized by emaciated, unhealthy bones, and an existence on the brink of survival. Underdevelopment in Italy has a longer history than the phases of 'economic miracles', even though it leaves fewer traces in the subsoil. The archaeologist's spade and trowel try to reconstruct the continuity of history through the long intervals of darkness.

[1980]

The Narrative of Trajan's Column

The metal network of scaffolding and planks which for some time have been wrapped round various Roman monuments offer a rare if not unique opportunity in the case of Trajan's Column. This is perhaps the first time in the nineteen centuries since the column was originally erected that there has been such an occasion: the chance to see the bas-reliefs from close up.

We are seeing them perhaps in a perilous condition, for the marble of the sculpted surface is turning to chalk, which dissolves in water, and the rain has been washing it away. The Department for Antiquities is trying to protect this thin, now crumbling layer with scaffolding, buying time while waiting for the discovery of a system to hold it in place; but we don't know if such a system exists yet. Whether it is the fault of the smog, of the vibrations, or just the effects of time, which, millennium after millennium, erodes everything to dust, the fact is that the presumed eternity of Roman remains has perhaps come to its twilight, and our fate will be to witness its end.

When I heard this, I rushed to climb the scaffolding around Trajan's Column, certainly the most extraordinary monument that Roman antiquity has left us, and also the least well known, despite the fact it has always been right in front of our eyes. For what makes the Column exceptional is not just its height, 40 metres, but the 'narrativity' of its figures (which is all about minute details of great beauty). The narrative requires a consecutive 'reading' of its spiral of reliefs, 200 metres in length, which tell the story of the two wars fought by Trajan in Dacia (AD 101–102 and 105). Accompanying me on this visit was Salvatore Settis, professor of classical archaeology at the University of Pisa.

The story begins by representing the situation immediately before the beginning of the campaign, when the Empire still ended at the Danube. The narrative strip opens (at first very low down then gradually rising upwards) with the landscape of a fortified Roman town on the river, with its walls, look-out towers and beacons in case of incursions by the Dacians: piles of wood for fires, piles of hay for columns of smoke. All elements that are meant to create a sense of alarm, of waiting, of danger, like in a John Ford western.

Thus the scene is set for the next relief: the Romans crossing the Danube on pontoon bridges to make a bridgehead on the other bank. Who can doubt the absolute necessity of reinforcing that border which was so exposed to barbarian attacks by establishing outposts in their territories? The ranks of soldiers walk over the bridges; at their head are the legions' standards; the figures evoke the clanking tramp of marching troops, with helmets dangling from their shoulders and mess-tins tied to poles.

The protagonist of the story is, of course, the Emperor Trajan himself, who is portrayed sixty times in these reliefs; one could say that each episode is marked by the reappearance of his image. But how does one distinguish the Emperor from the other characters? Neither his physical aspect nor his dress offer distinctive signs; it is his position in relation to the others that denotes him without any shadow of doubt. If there are three figures in togas, Trajan is the one in the middle: indeed, the two on either side look at him, and it is he who directs matters. If there is a row of people, Trajan is always first, or he is in the position of haranguing the crowd, or of accepting the submission of the conquered: he is always in the place where the gaze of other people converges, and his hands are raised in eloquent gestures. Here, for instance, you can see him ordering a fortification to be built, pointing to the legionary who is sticking his head up from a ditch (or from the waves of the river?) and carrying on his shoulders a basket full of earth from the excavations for the foundations. Further on he is portrayed against the background of a Roman camp (in the middle of it is the imperial tent) while the legionaries push a prisoner in front of him, holding him by his hair (the Dacians can be made out by their long hair and beards) and, using their knees (almost as if tripping him up), they force him to kneel at his feet.

Everything is very precise: the legionaries are distinguished by their ribbed breastplates (a piece of armour with horizontal ridges), and, since they also had to perform the duties of sappers, we see them building a wall with stones or chopping down trees still with their breastplates on – an unlikely detail but one which lets us know who they are – whereas the *auxilia* (auxiliaries), who are more lightly armed, and are often portrayed on horseback, wear a leather waistcoat. Then there are the mercenaries who come from the conquered peoples: they are bare-chested, armed with clubs and have features that suggest their exotic origin, including Moors from Mauretania. All the soldiers sculpted in the reliefs, thousands and thousands of them, have been catalogued with precision because Trajan's Column has hitherto been studied primarily as a document of military history.

More problematic is classifying the trees, which are represented in a simplified form, almost as ideograms, but still capable of being grouped into a restricted number of clearly distinct species. There is one kind of tree with oval leaves, and another with wispy leaves; then there are oaks, with their unmistakable foliage; and I think I can recognize a fig-tree as well, sticking out from a wall. Trees constitute the most frequently recurring landscape element, and often we see them falling beneath the axes of the Roman woodcutters, either to supply beams for fortifications or to clear the way for roads. The Roman advance opens up a path in the primeval forest just as the sculpted story opens up a passage in the block of marble.

As for the battle scenes, each of them is also different from the next, as in great epic poems. The sculptor has frozen them at the crucial point where the outcome is decided, arranging them according to a visual syntax that clearly stands out with great elegance and nobility of form: the fallen are at the bottom like a frieze of supine corpses at the edge of the 'strip'; then there is the movement of the two armies clashing, with the victors in the dominant position; above them again is the Emperor, and in the heavens a divine apparition.

And also just as in epic poems, they always have a macabre or violent detail: here is a Roman holding in his teeth the severed head of a Dacian enemy, the long-haired head dangling from his mouth; and other severed heads are presented to Trajan.

One gets the impression that every battle is also distinguished by a motif of geometric stylization that is different every time: for instance, here we see the Romans all with their right forearm raised at right angles, all in the same direction, as though throwing a javelin; and immediately above them is Jupiter, soaring with his robe like a sail, raising his right hand in exactly the same gesture, brandishing what was certainly a golden thunderbolt that has now disappeared (we are supposed to imagine these reliefs as coloured, just as they were originally), an unmistakable sign that the gods favoured the Romans.

The rout of the Dacians is not chaotic: instead, they maintain a mournful dignity even in their suffering. Away from the melee two Dacian soldiers are carrying a wounded or dead comrade: this is one of the most beautiful friezes on Trajan's Column, and perhaps of all Roman statuary, a detail that was surely the source of numerous Christian Depositions. A little above them, amidst the trees in a wood, the Dacian king Decebalus sadly contemplates the defeat of his men.

In the following scene a Roman with a blazing torch is setting fire to one of the Dacian cities. It is Trajan himself who is giving him the order to do so, standing there behind the soldier. Tongues of fire (we imagine them painted red) lick round the windows while the Dacians start to flee. We are just about to condemn the Roman conduct of the war as merciless when on closer examination we see sticking up from the walls of the Dacian city poles with severed heads stuck on them. Now we are ready to condemn the Dacians as cruel and to consider the Romans' revenge as justified: the orchestrator of the reliefs knew well how to balance the emotional power of the imagery with his pursuit of a celebratory strategy.

Then Trajan receives an embassy from his enemies. But by now we have learned to distinguish between the Dacians wearing a *pilleus* (a round cap), who are nobles, and those who have long hair and wear no headgear, in other words the ordinary people. Well, the embassy is made up of men with long hair, and that is why Trajan does not accept them (his gesture with three fingers is a sign of rejection); it is clear that he is demanding to speak to those at a higher level (which will soon happen after further Dacian defeats).

Suddenly we see an unusual sight in this story that is totally male dominated, like so many war films: a young woman with a look of desperation on a ship that is leaving a harbour. There is the crowd bidding her farewell from the jetty, and a woman holding out a baby boy towards the departing woman, no doubt a child of hers whom the mother has been forced to leave behind. Inevitably Trajan is here too, witnessing this farewell. The historical sources explain the significance of this scene: she is the sister of King Decebalus, who is being sent to Rome as war booty. The Emperor raises one hand to say farewell to his beautiful prisoner and with the other points to the boy: perhaps reminding her that he will hold the little boy as a hostage? Or promising her that he will have him educated in the Roman way in order to make him a subject king of the Empire? Whatever its significance, the scene has a mysterious pathos, heightened by the fact that in the same sequence, we're not sure why, we have just seen a raid on animals, with images of slaughtered lambs.

(Female figures appear also in one of the cruellest scenes on the Column: furious-looking women are torturing naked men – Romans, it seems, since they have short hair, but the significance of the scene remains obscure.)

The break between the scenes is marked by some vertical element, for instance a tree. But sometimes there is also a motif that continues over the break, from one episode to another, for instance the waves of the sea over which the prisoner princess sails away become the current of the river which in the following scene overwhelms the Dacians after their vain attempt to attack a Roman stronghold.

Alongside the horizontal continuity (or rather diagonal continuity since we are dealing with a spiral winding round a marble trunk) we notice motifs that are linked vertically from one scene to another up the height of the Column. For instance: the Dacians have along-side them the Roxolani, cavalrymen whose bodies are entirely cov-ered by armour made of bronze scales, and their horses too are all covered with these scales; their showy presence, almost a foretaste of medieval imagery, dominates in a battle by the river; but in the scene of another battle which comes immediately above this one we see another of these scaly creatures lying dead, stretched out like a kind of man-fish or reptile-man. Later on the movement of a battle is conveyed by ranks of oval shields advancing in a diagonal line; in the portion of column above it we see a series of shields of the same shape but this time deployed horizontally: they have been flung to the ground by enemies who have surrendered in another battle.

The spiral twists and follows both the development of the story in time and its journey through space, so the story never returns to the same place: here Trajan boards ship in a harbour, there he lands and starts marching to pursue his enemy, suddenly there is a fortress attacked by the Romans in *testudo* formation, and further on the field artillery enters the scene: the *carrobalistae* or catapults mounted on carts. Everywhere the frieze records the fallen and the wounded on both sides, as well as the medical care for which Trajan's army was famous. One can clearly see the effort that has been made not to play down the contribution from any of the Roman army's corps: if a wounded legionary is shown, by his side is placed another wounded man from the *auxilia*.

After the final battle of the first Dacian campaign Trajan is seen receiving the supplications of the defeated, one of whom embraces his knees. King Decebalus is there too amongst the suppliants, but set apart and more dignified. A winged Victory separates the end of the story of the first campaign from the beginning of the second, with Trajan setting sail from the harbour of Ancona. But for the moment this is as far the scaffolding goes, and I have not been able to see how it ends. I will tell you the rest of the story as soon as I have been able to see it for myself.

Finally we must mention the great mystery that surrounds this monument: a column so high and totally covered in scenes that have been sculpted in minute detail but cannot be seen from the ground. Of course, in the first century AD there were tall buildings around here that have now disappeared, whose terraces looked out on to the Column; but the distance from which these spectators had to observe it was not such as could allow a 'reading' of all the details, and in any case it was impossible to follow the continuation of the story along its spiral path. (This scaffolding is perhaps not too different from that used by the archaeologists sent by the sovereign heads of Europe to go up and make their drawings and casts: François I, Louis XIV, Napoleon III, Queen Victoria. More adventurously Ranuccio Bianchi Bandinelli had himself hoisted up there on a firefighters' ladder. These explorations, however incomplete and irregular they were, have been carried out from one century to another, and it is thanks to their results that we have been able to study Trajan's Column up till now.)

It is not only the addressee of this elaborate visual message that remains a mystery. We know nothing of the system whereby the eighteen 'rocchi' (or cylindrical marble blocks, hollow inside and with a spiral staircase in the centre), which make up the Column's shaft, were hoisted on top of each other. Nor do we know whether these 'rocchi' were sculpted on the ground one by one or only after they had been raised up into place.

Then there are other mysteries: how could the ashes of Trajan and his wife have been walled into the base of the Column if a mandatory Roman law forbade the burial of the dead inside the *pomerium* (city precincts)? (Those collected in a golden urn were not his real ashes, but it was as if they were: Trajan, who had died at Selinunte and was cremated there, was represented at his triumph in Rome by a wax model, which was later burned with the honours due to an Emperor who was destined to ascend into heaven.)

On the other hand, the major interests that the Roman conquests in the Black Sea area entailed (Dacia was rich amongst other things in goldmines) fully explain the grandiose nature of the cult of Trajan (the celebration feasts lasted 180 days; the donation that each citizen received was the most generous ever recorded) and the complex of gigantic monuments around the Emperor's tomb and temple. What remains for us all the way down to our own times is this epic in stone, one of the most copious and perfect visual narratives in history.

[1981]

The Written City: Inscriptions and Graffiti

When we think of a Roman city in imperial times we think of temple colonnades, triumphal arches, baths, circuses, theatres, equestrian monuments, busts and herms, bas-reliefs. It does not occur to us that this silent scenery made of stone lacks the most characteristic element, even from a visual point of view, of Latin culture: writing. The Roman city was above all a written city, covered by a layer of writing that went across pediments, tombstones, shop-fronts. Armando Petrucci writes:

Writing was present everywhere, painted, scratched on to surfaces, engraved, placed on wooden tablets or traced on to white squares . . . sometimes advertisements, sometimes political graffiti, sometimes to do with funerals, with celebrations, now public, at other times very private, notices or insults, or good-humoured memories . . . displayed everywhere, with a preference, it is true, for some specially chosen spots such as squares, fora, public buildings, or necropolises, but these were only for the most solemn forms of inscription; not like other writing which was scattered all over the place wherever there was a shop-entry, a crossroads, a piece of blank stucco at human height.

However, in the medieval city writing disappeared: both because the alphabet had ceased to be a medium of communication within everyone's reach and because there were no more spaces available to accommodate writing or to attract people's eyes. The roads were narrow and winding, the walls all protuberances and bumps with ornamental mouldings under arches; the place where all discourses about the world were transmitted and kept was the church, whose messages were oral or figurative rather than written.

These two opposing images are suggested by Armando Petrucci at the opening of his article 'La scrittura fra ideologia e rappresentazione' (Writing between Ideology and Representation), which – in 114 pages and 122 illustrations – constitutes the first historical outline ever of inscriptions in Italy from the Middle Ages to today, and not only of inscriptions but of every example of visible writing and thus, in short, of what today we call graphics. *Grafica e immagine* (*Graphics and Images*) is in fact the title of the new volume of the *Storia dell'arte italiana* (*The History of Italian Art*), published by Einaudi (Part III, volume 2, tome 1), of which Petrucci's is one of the chapters.

In the medieval city Roman inscriptions continued to speak with their own solemn voice that few now understood. At the same time the tradition of writing perfectly executed characters was preserved in the pages of manuscripts written inside cells by monks who were scribes, using techniques and models that were by now completely different. As a result, when from 1000 onwards words were needed for the walls of cathedrals and palaces, there would be two letter-forms they would use for their inscriptions in faulty Latin, either as alternatives or in combination: straight block capitals, as in ancient inscriptions, or the alphabet they found in books, which was Gothic, spiky and twisted, and which filled the walls thickly as though they were pages.

Nothing seems more static and codified than Latin capitals. And yet it is precisely when the Roman letter-form comes back to prominence in the fifteenth century that the adventures of each letter can be followed in the restricted range of whimsical ornamentation they developed. The letter Q is the one that allows itself most whims, since its most characteristic feature is its ability to wag its tail as it wants: a cat-letter that like a feline curls round itself and moves its tail, now lengthening it under the following letter, now hurling it in lightning-sharp whiplashes, now dragging it lazily and making it curve in either convex or concave undulations. But A too can afford some liberties, for instance resting all its weight on its left leg, or (in less orthodox variants) bending its bar at an angle, while M can choose between a position of being at ease, with its legs spread wide, or one of attention with its legs vertical and parallel. The G can end with a rounded curl or with a sharp tooth or with a pug-nosed hook, or close in on itself like an alembic. The letter X can escape its arithmetical and algebraic vocation by varying the angles of its crossing or allowing one arm to stretch out in undulating movements. As for Y, it never misses an opportunity to stress its non-Latin origin by adopting the form of a palm-tree with curved leaves. Sometimes conventions of epigraphic abbreviation prompt the invention of new signs, such as an NT which is condensed into one ideogram, a letter which itself acts like a bridge and not by chance appears in the plaque celebrating the construction of a bridge dedicated to a 'pontiff' (the Ponte Sisto, 1475).

Initially determined by the act of engraving with a chisel or writing with a pen, the shape of alphabetic characters quickly adapted to the needs of the new art of printing, which soon held sway over all types of writing. And printed frontispieces taught a new sense of proportions, of relationships between white spaces and black characters, which was immediately picked up in stones and plaques. The composition of pages in print soon produced bizarre, spectacular paginations, as in the *Hypnerotomachia Poliphili* by Francesco Colonna, a book printed in Venice but conceived in Rome.

Almost all of these developments in this history of graphic visibility take place in Rome, in the sight of Roman remains and in dialogue with them. After Michelangelo, who plays an important role in this dialogue – his role being halfway between a renewal of the classical order and innovation – the Baroque revolution starts to break out. The pleasure of fiction starts to gain the upper hand, and it is no longer so much the writing that counts as its material support, which deforms it and sometimes hides it between drapes and linings. What we find are commemorative stones and plaques, in bronze or in black or red marble, in the shape of scrolls or drapes or shrouds or animal-skins, surfaces that are either in movement or crumpled or torn at the edges, where metallic or golden letters wave and disappear between the folds. Just as stonework pretends to be a page, so in the frontispieces of books the page pretends to be a stone. Thus we arrive at Piranesi, in the visionary and eclectic eighteenth century which runs alongside and counterbalances the neoclassical and purist eighteenth century of Bodoni and Canova.

When we come to the modern era, Petrucci stops following the dominant line of graphic taste, which was becoming less interesting artistically, to try to catalogue the 'departures from the norm'. From this point of view he starts his story from scratch again, exploring scrolls by the Sienese primitive artists, as well as astrological charts, guild-emblems, and ex-votos. The fantastical forms of popular graphics are a spontaneous vegetation that will be cultivated and harvested by the avant-gardes, starting with William Morris, who will proclaim the revolution against Bodoni.

In a rapid sketch he brings us all the way down to 1930s Italy, where, in a nod to modernity, the most simple and austere character, the unadorned sans-serif, is adopted as the official font of the Fascist regime, which thus transformed the functional lines of Bauhaus design into something more authoritarian and neoclassical. As for recent times, contrasting with this picture is not so much a left-wing graphic style (though here Petrucci gives prominence to the 'losing side' and traces a fine portrait of Albe Steiner) as the illegal explosion of graffiti on walls all in support of current protests.

It is thus right that Petrucci's article ends on this invasion of script 'from below', characterized by an 'anti-aesthetic' urge. This anti-aesthetic impulse is the most glaring aspect of the protest by the young and the excluded in society, a protest which has been going on now for some twelve years, starting, of course, with the famous slogans from May '68 in Paris and from the phenomenon of the tag 'signatures' on the New York subway (a phenomenon with particular characteristics that are more linked to artistic intentions).

The 'palimpsests' that these illegal writings form, as they are superimposed on previous 'official' inscriptions of all kinds, which act as a simple 'support' surface, or as they become entangled with later interventions by militants of opposing factions, become in Petrucci's study a precocious object of study analysed by a method that is almost palaeographical. However, the technical objectivity of Petrucci the scholar does not hide the sympathetic attitude he displays for this graphic jungle, where he recognizes a 'growing importance of writing as a semantic instrument and as an aesthetic product in the urban space'. This does not prevent him recording also the degradation caused by such urges, which we witness in writing that has nothing behind it except an ill-defined and lazy arrogance, writing which so frequently occupies the walls of Italian cities. This historical survey ends significantly with the desolate vision of the Foro Italico, where the letters of Fascist inscriptional rhetoric mix with the violent graphic screams of the fanatics who support football teams.

Now that I've reached this point, now that I have done my duty as regards information and summarized the content of this essay in all its richness and sophistication, it is time to come out with the objection I have been holding back from the start. From the first page, when he evokes the city of Rome totally covered in writing, both official and private, down to the last, where he celebrates the guerrilla warfare of 1968 graffiti, Petrucci pursues his ideal of the 'written city', a place saturated in messages which are structured using alphabetical signs, a place that lives and communicates through the positioning of words that can be exposed to people's gaze. Now that is precisely the ideal I disagree with. Words on walls are words imposed by someone's will, whether that person is high up or low down, words imposed on the gaze of all the others who have no choice but to see them or receive them. The city is always a transmission of messages, it is always a discourse, but it is one thing if this is a discourse that you have to interpret yourself and translate into thoughts and words, and quite another if these words are imposed on you without any chance of escape. Whether it is an inscription celebrating authority or a defamatory insult, we are still dealing with words that land on us at a point in time which we have not chosen: and this is a form of aggression, abuse, violence.

(Of course the same applies to the writing produced by advertising; but there the message is less intimidating and conditioning – I have never believed much in 'hidden persuaders' – it finds us more prepared and it is in any case neutralized by the thousands of equally powerful competing messages.)

The written word is not an imposition if it comes to you through a book or a newspaper, because in order to be received it presupposes a previous act of consent on your part, an agreement to listen which was expressed in your buying or just in opening that book or paper. But if it comes to you via a wall which one has no chance of avoiding, then it is a form of tyranny however you look at it.

There are people today who feel the need to assert that their rights have been trampled on by writing about them on walls with a spray-gun. The day they have power they will continue to need walls to justify themselves, using bronze or marble letters or – depending on the customs of the time – huge propaganda banners or other tools for brainwashing people.

This discourse of mine does not apply to graffiti under oppressive regimes, because there it is the absence of free speech that is the dominant element even in the visual aspect of the city, and the clandestine writer fills this silence entirely at his own risk: even reading it is in some sense a risk, and imposes a moral choice on us. Similarly I would also make exceptions to my rule of thumb for cases where the writing is witty, as we have often seen recently, both in Paris and in Italy, or when it is such as to prompt an illuminating reflection or poetic evocation, or uses its graphic form to portray something original. To see the value of this humorous or poetic or aesthetically visual thought involves an operation that is not passive, an interpretation or decoding, in short a collaboration on the part of the receiver who appropriates it through some mental effort, however instantaneous. But where the writing is simply a naked affirmation or negation which requires from the receiver merely an act of consent or refusal, the impact of being coerced into reading in this way drowns out any potential advantage that comes from managing to re-establish our internal freedom in the face of verbal aggression. Everything is lost amidst the din of the neuro-ideological bombardment to which our brains are subjected from morning to night.

Nor would I feel like taking the cities of the Roman Empire as a model, where all the official written and architectural messages were imposed by imperial power and state religion. If today Roman writing attracts us it is because its messages require on our part a decoding system that is to some extent a dialogue, freely participated in: its intimidatory power is now extinct. In the same way, the function of Arabic script in architecture and in the whole visual world of Islam seems to us full of fascination: we notice the presence of the written word, which envelops its spaces in an atmosphere of thoughtful calm, but we are safe from the power of injunction in that script because we cannot read it, or – even if we do know how to read it – because it seems distant from us, sealed shut in its formulae. (The same applies to the calligrams of the Far East.) It is the presence of writing, the potential of its varied and continual uses that the city has to transmit, not the abuse of power in its actual manifestations. Perhaps this is the point where Petrucci's thesis and my argument meet up: the ideal city is the one over which hovers a dust-cloud of writing that does not calcify or turn into sediment.

But have not the poor walls of Italy's cities also now become a series of layers of arabesques and ideograms and hieroglyphics superimposed on each other, so much so that they no longer transmit any message except that of dissatisfaction with every word and our regret at this wasted energy? Perhaps writing finds a place that is uniquely its own on these walls too, when it refuses to be abused by arrogance and tyranny: a noise which you have to strain your ear carefully and patiently for until you can make out the rare, discreet sound of a word that is for a moment true.

[1980]

Thinking the City:
The Measure of Spaces

Around the year AD 1000 Europe experienced an urban development of a kind that it had not seen since antiquity. The medieval city that had taken shape over the previous four centuries showed profound differences from the ancient one from which it had very often inherited its site, name and even its very stones: all the structures linked to the social life of the past had disappeared (temples, forum, baths, theatres, circus, stadium). Its geometric structure too based on the two great perpendicular axes was no longer recognizable, obliterated as it was by labyrinths of narrow, winding streets; the churches, the principal reference points in the Christian city, were distributed irregularly, in sites connected with the lives of the saints, miracles, martyrdoms and relics.

It was the network of churches that shaped the city, not vice versa, as did the hierarchy that was established amongst them: the cathedral, which was the bishop's see, would be the religious and social centre; but the city had as many centres as it had parishes, plus the convents of the various orders; the routes of processions would determine the importance of the city's various arteries.

The medieval city was the city of the living and the dead: corpses were no longer considered impure and relegated outside the circle of city walls; familiarity with the dead and contact with the necropolis were one of the great transformations of urban culture.

The straight lines that the city's horizontal dimension had lost resurfaced instead in the new vertical dimension: the city of church-towers emerged (from the seventh century onwards), where the chimes from on high counted out the hours and confirmed for the Church its 'dominion over time and space', and then the city of civic towers developed, rising beside the town hall and the barons' residences, as soon as civic power established itself (from the thirteenth century onwards) alongside the ecclesiastical authorities.

It was the function of the city that had changed: it was no longer a military and administrative space as it had been in the times of the Roman Empire, but a city of production and exchange and consumption. The market was more and more in the hands of the city's most representative class, the bourgeoisie.

Compared with other European cities of the time, Italian cities were characterized by a much heavier presence of Roman antiquities, by signs of the predominance of the Germanic Emperors, or of the resistance to their descents into Italy (for instance, citadels and fortresses), by the presence of an urban aristocracy that was no longer holed up in its castles, by being surrounded by a countryside that was subject to the town, and by the independence of the city-states.

I am summarizing an essay by Jacques Le Goff, on 'L'immaginario urbano nell'Italia medievale (secc. V–XV)' (The Image of the City in Medieval Italy (Fifth to Fifteenth Centuries)), which dwells in particular on texts from a literary genre typical of the time, the *Laudes Civitatum* (City Eulogies): the most famous is that of Bonvesin de la Riva in praise of Milan. Le Goff traces the real or imaginary models in relation to which Italian cities were seen or thought about by their inhabitants, for instance, comparisons with Jerusalem – the earthly or heavenly one – or with Rome. (The article opens the fifth volume of the *Annali* of the Einaudi *Storia d'Italia* (*History of Italy*), which is entitled *Il paesaggio* (The Landscape), and is edited by Cesare De Seta.)

A passage from Leopardi could be taken as emblematic of the relationship between real places and our way of thinking about them or experiencing them. (It is quoted by Sergio Romagnoli in another fine essay in the volume, on landscape in Italian literature from Parini to Gadda.) In the early days of his stay in Rome (December 1822), Leopardi writes to his sister Paolina that what has struck him most is the disproportion between human dimensions and the size of buildings and spaces: the latter would be fine 'if men here were five arms high and two wide'. What causes him anguish is not just the emptiness of St Peter's Square, which the population of Rome is not enough to fill, or the mass of the huge cupola, which, when he sees it on arrival, seems as high as the Appenine peaks. Instead, it is the fact that 'all the grandeur of Rome serves no other purpose than to multiply distances, and also the number of steps that one has to climb up to see whoever it is one wants to see ... I don't mean to say that Rome seems uninhabited to me; but I do say that if men felt the need to live in such an expansive way, as one lives in these palaces, and as one walks in these streets, piazzas and churches, the whole globe would not be enough to contain the human race.'

This is a sensation that differs considerably not only from our experience of our age of over-population but from the experience of European capitals that were crowded and tumultuous, which was what writers like Fielding and Restif de la Bretonne had experienced, and what, soon after Leopardi, Balzac, Dickens and Baudelaire would come to know. Leopardi's agoraphobic vision puts us into a dimension of city landscapes dominated by emptiness which can really be said to be a mental constant in Italy and which connects the images of 'ideal cities' from the Renaissance with the metaphysical cities of De Chirico.

In order to convey this sensation Leopardi invites Paolina to think of a chessboard as big as the main square in Recanati, with chesspieces of normal size moving on it. From the first evocation of a city of giants to that of a city of dwarves: Leopardi's imagination hovers between Brobdingnag and Lilliput, as Sergio Romagnoli notes.

A few days later, writing to his brother Carlo, Giacomo establishes his idea of the 'sphere of relations' between men and things, such as can be realized in small environments, in small cities, but are lost in big ones. Here we touch on a crucial focus in Leopardi's poetry: the relationship between a confined, reassuring space and a beyond that is boundless and inhuman. On one side there is the house, the window, the familiar evening noises of Recanati, its 'lanes bathed in gold sunlight and the orchards'; on the other stands the immensity and indifference of Nature as she appears to the

Icelander in one of his *Operette morali*; on one side the hedge of
'L'infinito' and on the other infinity. This is a contrast in which repul-
sion and fascination can swap sides: his native town, a model of
human dimensions, is also unbearable; and drowning in the sea of
the boundless void can be sweet. As for the theme of the Italian
landscape, Sergio Romagnoli places in contrast to these Leopardian
themes the idealization of the small town in German Romanticism.

Not many years before this an eccentric German, Johann Gott-
fried Seume, had set out to discover what he called 'real Italy' and
this he identified as small-town Italy: scorning diligences and car-
riages and itineraries devoted solely to monuments, he went everywhere on foot (travelling 30 kilometres a day). The aristocratic and
humanist tradition of the Grand Tour in Italy comes to an end with
Seume, who reverses its rules. So says Cesare De Seta, who devotes
a lengthy essay to this vitally important experience in the history of
European culture.

The journey through Italian towns that the educated and
wealthy (French, British, German) foreigner was required to complete underwent various changes between the end of the sixteenth
and the end of the eighteenth centuries: there are locations that
appear and disappear, others that change in importance. De Seta
has studied travel-diaries in order to compare and interpret these
changes in perspective. In the end, after the Napoleonic wars, the
epoch of the Grand Tour comes to a close and the age of tourism
begins, in a Europe where the distances between nations continually shrink.

Amidst the other articles in the book illustrating the idea of Italy
as an image, two are on topics that are likely to arouse an ironic reaction in Italians. One is on guidebooks, Baedeker and the Touring
Club Italiano (by Leonardo Di Mauro); the other is on the stereotypical images of towns such as are found on picture postcards (by
Maria Antonietta Fusco). But I notice with relief that the Touring
Club guidebooks – which are one of my secret passions and I believe
one of the things that newly unified Italy knew how to do well – are
treated with the respect and *pietas* they deserve, even as regards their
weaknesses, lacunae and clichés.

As for stereotypes, such as the image with the pine in the fore-ground and Vesuvius as backdrop, our reactions are inevitably sarcastic. But perhaps we should not just see in such images a product of 'mass culture': a country starts to be present in people's memory when every place-name has an image connected to it, an image which as such does not mean anything other than that name, an image which is as arbitrary or justifiable as any name. Leaning Towers and Turin's Mole Antonelliana are nothing but concise iconic abbreviations, or coats of arms, or allegories. The important thing is that they serve to distinguish, not to confuse or flatten, differences, unlike the Venetian gondolier singing the Neapolitan song 'O sole mio' in Ernst Lubitsch's film *Trouble in Paradise* – though it has to be said that this incongruous splicing together of two stereotypes undoubtedly has some semantic relevance for signifying tourist Italy, and in addition actually reflects the reality of consumer tourism when it comes into contact with gondolas and music in the Italy of today.

[1982]

The Redemption of Objects

The personal anthology that Mario Praz has compiled – *Voce dietro la scena* (Voices Off), published by Adelphi – brings together essays and chapters from his oeuvre written over the course of more than fifty-five years. One of the main themes is that of autobiography. Praz was a scholar with an insatiable appetite for learning about and comparing things; he was an omnivorous compiler of files about great, minor and negligible works where the human hand has expressed the unmistakable tone of the age as well as the hidden urges of the soul. He was an explorer of the remotest sources of the currents of taste that have irrigated the entirety of Western culture. As a historian of taste Praz does not proceed in linear fashion in his work but through a juxtaposition of materials in which every element refers in turn to other series of elements. Similarly this autobiography is not an ordered account of events in chronological order but an accumulation of motifs and opportunities and stimuli, or rather the catalogue of the rationales that have underpinned and given shape to his life.

Thus we find here his vocation to be an English literature scholar identified at its source, in his study of English eccentrics who once stayed in Italy, especially in Tuscany (such as Vernon Lee, writer and disciple of Ruskin, and William Morris, a curious character who combined Pre-Raphaelite aestheticism with Tolstoyan humanitarianism). It is also seen in his scouring of London in search of places described by Charles Lamb, whose essays he had just translated (it was Papini who had commissioned this first work from him for his famous series 'Cultura dell'anima' (Culture of the Soul) in 1924), and in the years spent teaching in Liverpool, impatiently trying to squeeze out of the dullness of the modern industrial city every last drop of fascination for the ancient civilization that was the only thing that attracted him.

Thus we see that his survey of the origins of Decadentism (or rather of the Romanticism-Decadentism nexus), which would be the theme of his first and most famous book, *La carne, la morte e il diavolo* (*The Romantic Agony*, 1930), stemmed from his research on D'Annunzio's predecessors and sources, as well as from a trip to Spain and from his reflections on the bullfight in literature. His interest in Mannerism, which is a related area, came from his love of Tasso (Tasso whom we see in one essay unexpectedly placed alongside Diderot: one of the pleasures which a reading of Praz always holds is just such unpredictable juxtapositions, his short-circuiting of thematic and stylistic analogies). Then we have his passion for collecting: Empire furniture, for instance, where he records his first purchases with the meagre savings he had put aside as a student; paintings of interiors which, even when they were not top-class works, have so much to tell us, as part history of taste and part narrative; and wax images where the suggestion of the living person and his ghostly appearance are both equally present.

This relationship with objects is another key part – the most essential one, I believe – of this 'personal anthology' by Praz: it can also be found in other typical books of his, from *Il gusto neoclassico* (*On Neoclassicism*) to *La casa della vita* (*The House of Life*). Indeed it is in this relationship that what we can call Praz's philosophy is defined. Two essays in this volume illuminate this philosophy in particular, 'Dello stile Impero' (On Empire Style) and 'Un interno' (An Interior). Both were written to defend Empire furniture from the accusation of being lugubrious and sinister – an accusation documented in a great number of literary references which Praz, suffering all the while, collects together, and up to a point he takes pleasure in emphasizing the effects that might support these people, his enemies. However, he also advances his own defence to a certain extent, but doing so as though he is someone who knows that it will not be understood, that it will be a secret that is difficult to communicate to others. It is the secret of a man who has managed to find 'in the sheer repetition of certain decorative motifs . . . an almost magical atmosphere . . . which comes together as a solemn calmness'. 'Did not sphinxes, chimaeras and other fantastic creatures find their raison d'être in a quintessence of nature, in a nature relived in the human imagination, and reassembled according to the logic of a dream . . . ?'

In 'An Interior' we are told about the occasion when Emilio Cecchi came to visit, while Praz was still living in via Giulia, in Rome, in the spacious but gloomy apartment in Palazzo Ricci. 'Not knowing whether to be amused or worried', Cecchi asked Praz how he managed to live amidst such disturbing furniture. And Praz in turn had fun describing the house room by room, loading it with ghostly, dismal penumbras; only to then go through it once more in full sunlight and to exalt it in all its colourfulness, and to prove that 'the soul of neoclassicism is noble, serene and – whatever its detractors say about it – profoundly cheerful'.

Nevertheless, the point I wanted to stress is a different one: Praz realized that Cecchi's objection was not so much about his taste as about his possession of such furniture. 'Expensive beauty was repugnant to Cecchi . . . as decoration for his own house he loved objects in which a minimum of intrinsic value was combined with a maximum of expressivity . . . Modest things, things that aided devotion and nothing else, yet devotion' – and he would emphasize this point – 'has to be totally spiritual, disinterested, uncontaminated by *the crude love of possession.*'

This is the controversial point which sees the ascetic and the collector face to face as in a work of ethics or a philosophical dialogue. On the one hand: 'This business of possession was a heresy for him: out of Cecchi's mouth I would hear Tagore's condemnation against "foolish pride in furniture" being repeated'; on the other, 'This asceticism, as I said, is totally alien to me. I would not hesitate to repeat to my friend my brazen confession that I am a materialist, so that for me the sensual presence of things has enormous importance.'

This dispute had flared up several times before in the preceding essays of this anthology, so much so that one could almost say it is its *leitmotiv*. And the role of supporter of asceticism had been taken on in turn by Vernon Lee (in the first essay in the volume), that apostle of aestheticism and of rejection of possessions, then by Rabindranath Tagore, in the essay 'Dello stile Impero'. The Indian poet, in a lecture given in Florence,

singled out amongst the deplorable vices of the West 'the foolish pride in furniture'. In fact it seems absurd that one should be proud of an elegant little table or of a chair in a certain style or of a pair of candelabra: what good does it do to decorate a house till it becomes beautiful, when the human spirit, according to philosophers and poets, can still proceed as supreme ruler amidst four poor walls. Diogenes' barrel should be enough to provide protection for us human worms born, as Dante says, to form the angelic butterfly.

Then suddenly Praz rushes to marshal the opposing argument:

But immediately a doubt arises. Because such is the nature of these dear material things amidst which we live our lives that you can't deny one of them without denying all of them at the same time. To have set my soul on a little table or chair that has caught my eye is a sin that is only slightly worse than setting my soul on a landscape . . .

And yet the contemplation of natural landscapes passes for being the most spiritual thing possible: so why then is the contemplation of furniture not the same, especially as 'furniture obeys a law of economics which is the same as that which controls landscape'? These are passages from the 1930s, and it is no accident that an echo of Bauhaus theories is recognizable even in an author who is so far from them, someone who is totally concerned with the forms of the past. Furniture 'has artificial but not arbitrary forms; it has a rule of necessity that is the same as that which governs the mountains and the plains; and its beauty is in proportion to the extent to which it conforms to that rule'.

With the calm tone of someone who wants to examine the question from every angle, but always with a hint of sarcasm, beneath which the tenacity of his passion is visible, Praz affirms what he calls his 'materialism', in other words the rejection of any spiritual asceticism ('the truth is that I have a soft spot for fine furniture but no soft spot for Rabindranath Tagore'), but also the rejection of any reduction of the human to the bare nature of a biological or vitalistic or existential or psychological or merely economic entity.

The human is the trace that man leaves in things, it is the work, whether it is a famous masterpiece or the anonymous product of one particular epoch. It is the continuous dissemination of works and objects and signs that makes a civilization the habitat of our species, its second nature. If we deny this sphere of signs that surrounds us with its thick dust-cloud, man cannot survive. And again: every man is man-plus-things, he is a man inasmuch as he recognizes himself in a number of things, he recognizes the human that has been in things, the self that has taken shape in things.

Here the philosophy that I have tried to extrapolate slips from the universal to the particular, or rather to the private, because it is here that the logic of collecting clicks into place, collecting that restores unity and a sense of the whole that is homogeneous with the dispersal of things. And the mechanism of possession also clicks into place (or at least the logic of the desire for possession): this mechanism is always latent in the relationship between man and objects, a relationship which, however, does not exhaust itself in it because its aim is the identification, the recognizing of oneself in the object. And in order to achieve this aim, possession clearly helps because it allows prolonged observation, contemplation, a symbiosis, a living together with the object. (But Praz, who follows the traces of his beloved objects also in books, in the non-corporeal existence of written texts, who becomes a collector of quotations, allusions, references, Praz is the proof of how much this concrete passion of his feeds on immaterial things.)

The identification of man with objects works in both directions, because the object does not just play a passive role. The collector:

through constant practice manages to see an antique shop from the other side of the street, and to note the authentic pieces which *call on him out loud* amidst the junk and imitations. How satisfying it is to redeem a good object in all its purity from the contamination of low and degrading company! I have often heard it said that if those pieces of furniture could speak, one could hear them pouring out their gratitude into our ears. The bookcase would fling open its glass doors impatiently to receive the volumes that are worthy to be on its shelves, the armchair would hug you in its embrace, the desk would stretch itself out to offer fresh inspiration to your pen. I am convinced, leaving fantasy aside, that furniture feels better physically – I was just about to say spiritually – when it is placed in its proper environment.

This is a passage from the article 'Vecchi collezionisti' (Old Collectors), which rightly belongs in any essential anthology of Praz the writer, or rather Praz the narrator. I would just add two other excerpts, two apparitions who are similar in their pathos: Charles V, now old and infirm, in the convent in Estremadura, roaming around amidst the ticking of his collection of clocks, and Mazarin, deposed and in exile, wandering round his collection of paintings, at night, in his gallery, bidding them farewell. This amorous relationship with things has at its base a layer of melancholy: just to give the last word to those who support the ascetic life.

[1981]

Light in Our Eyes

Every now and again I start making a list of the latest books I have read and of those I intend to read (my life functions on the basis of lists: accounts of things I have not completed, plans that are never realized). I read a medieval Persian poem, Nezami's *The Seven Princesses*, now translated into Italian, where each of the seven colours corresponds to a separate allegorical and moral field; then I read the Japanese writer Tanizaki's *In Praise of Shadows*, which talks of 'infinite gradations of darkness'; of course I have read Wittgenstein's *Remarks on Colours* (also recently translated): for him colours can be defined solely on the level of language; and that book spurred me on to reread Goethe's *Theory of Colours*, recently reprinted.

However, before all these books I had read another one which I had immediately wanted to write about, but I have held back until now, as happens with books which have too many interesting things in them to discuss properly in an article. Suddenly all my other readings in this area came to be connected with this one book, which tells us for instance that Newton, the man who discovered the spectrum through refraction, established that there are seven basic colours, not because he really saw seven colours, but because seven was the key number for the harmony of the cosmos (the seven notes in a musical scale etc.) and moreover he trusted an assistant who had such a discerning eye that he managed to distinguish a separate colour between dark blue and violet: this was indigo, which has a beautiful name but actually is a colour which has never existed.

Well, I cannot hold back any longer; I need to talk to you about this book: Ruggero Pierantoni, *L'occhio e l'idea: Fisiologia e storia della visione* (The Eye and the Idea: Physiology and History of Vision), published by Boringhieri. It is a history of the theories through which we have tried to understand how the eye works, what sight really is, what the nature of light is, starting with the Greeks and the Arabs and then coming all the way down to the modern age: it covers physiology, the philosophical premises behind every theory, and the consequences that stem from these for the arts, especially painting. I see from the blurb that the author 'has specialized in the biophysical aspects of communication in animals, working at the Max Planck Institute in Tübingen and the California Institute of Technology, and is currently a researcher at the Cybernetics Institute at Camogli, run by the Italian Centre for National Research'. A scientist with all the necessary qualifications, then, but one who cultivates an elegant style like that of a literary essayist, and has interests in the history of ideas and aesthetics as well as in the history of science and practical research.

There is a borderline territory between vision theory and problems in the figurative arts, an area where Gombrich's most famous books are situated; Pierantoni's book, especially in its final chapters, sails a course parallel to Gombrich's and in dialogue with him. But I am going to concentrate on the first three chapters, which are entitled: 'Myths of Vision', 'Space, Inside and Outside', 'Light, Inside and Outside'.

Pythagoras and Euclid believed that the eye emitted a set of rays which struck against objects; just like a blind man who advances stretching out his stick, so the seeing eye becomes aware of reality by touching it with its rays, which then return to the eye and inform it of what they have seen. Democritus thought that immaterial images detached themselves from things and then entered the pupil; but for Lucretius these were tiny fragments of matter, which he called atoms (and we call them photons). For Plato there were rays that departed from the eye and rays departing from the sun; they combined when they struck objects and were sent back to the eye. For Galen there was a visual spirit which originated in the brain, flowed into the eye, captured the light and images contained in the lens, and carried them back up to the brain.

Heirs of Greek science, the Arabs started from Galen, accepting the mediation of the visual spirit, but totally rejecting the idea of rays projected from the eyes towards the outside: by this stage vision came from outside, not from within.

The conviction that the eye emits light hit a crisis also in the Christian Middle Ages. It was in the lens (situated, despite what all experience tells us, in the centre of the eye, just like the Earth at the centre of the cosmos) that the fusion between the World and the Self took place: that was what Dante believed. The diagrams of the anatomy of the eye lost all reference to biology, acquiring a geometry of concentric circles just like – in Pierantoni's words – 'a Ptolemaic world of armillary spheres'.

By the time we get to the epoch of Leon Battista Alberti the rays departing from the eye have become geometric lines, Euclidean abstractions: the perspective pyramid. Then Leonardo dismantles this abstract construction: the 'property of sight' is not dot-shaped as it would be if it operated at the apex of the pyramid of lines, but is a property of the whole eye.

Leonardo's reflections on optics were inspired at times by his ingenious ability to stick close to reality, ignoring schematic patterns, and at other times by his efforts to make experience match the traditions that he had learned about in books. He was the first to realize that the optic nerve could not be a hollow canal, as antiquity and the Arab and Christian Middle Ages had believed, but something multiple and complex, otherwise the images would end up being superimposed and getting confused with each other. Meanwhile, in his paintings, it is the physiological, not the conceptual, nature of vision that he tries to capture.

For Leonardo light was never an abstract ray moving in the mind and eye of man, but a sea of rays constantly interacting with matter. And matter, objects, men, towns, were not representable through the continuous precise lines of their contours but could only be evoked by the constant fading of surfaces.

Meanwhile official science saw Vesalius publish his illustrations, in which anatomy had become an experience-based science founded on the dissection of corpses. Except the eye, however, which continued to be drawn according to traditional Graeco-Arab schemes. Leonardo's ingenious hypotheses stayed buried in his private jottings.

In the Italian painters of the Renaissance 'light is so omnipresent as to seem absent, and it does not appear to come from any point in the universe': it is a sea in which the figures are immersed. In the North, on the other hand, the idea of light is completely different:

The Flemish and Dutch learned to love those things where light is caught, imprisoned in a network of reflections, and from which it emerges transformed into rainbows. Enamel, crystal, steel, coral, quartz. Out of this comes a science that pursues and surprises light at the critical points of its journey through matter and in the closed secret of the human eye.

This was true, though with many differences from painter to painter:

Van Eyck paints things as he knows they must be, and Vermeer as he perceives them. In Vermeer light is a subjective, private thing . . . In the miraculous hands of Van Eyck it is the absolute revelation of a spiritual world that is destined for the eye of the soul and emitted by the eye of God.

Starting from antiquity and the Middle Ages, the metaphors which acted as models for the functioning of the eye changed many times: the stick, the arrow, the lens, the pyramid, then (in Leonardo's time) the camera obscura, then 'the mirror of the world', and the 'window of the soul'. When in 1619 Scheiner cut the white of the eye, looked 'inside' the eye, and saw 'as it were from a window' the image in the retina 'reflected as in a mirror', these two metaphors became decisive. Painters began to paint a little window reflected in the pupil of faces in portraits; Dürer's hare too, hidden in the grass, has a window in its attentive pupil.

As for the mirror, Claude Lorrain painted with his back to the landscape, which he saw reflected in a little convex mirror, conjuring up effects of distant, hazy beauty. The pathos of distance thus comes into being, a fundamental component of our culture.

The image reaches the retina reversed. How is it righted again?

Leonardo had hypothesized a supplementary lens in the eye's camera obscura, according to a system that was impeccable in optical terms, but devoid of any foundation in anatomy. It was Kepler who overcame the obstacle when he realized that turning the image the right way was an intellectual, not a physiological, operation. The time was ripe for the thinking and non-material *ego* of Descartes to enter the fray. But Descartes still needed anatomical backing for this, and he would choose the pineal gland, which was buried at the bottom of the brain, a well-defended fortress (the image is Pierantoni's) that guaranteed the unity of vision and subject.

By the way, though, why on earth do we need two eyes if vision is something singular (and the world is one)? The discovery of the chiasm (the meeting point of the two optic nerves) and, gradually, of its role and how it works brought philosophy into the picture.

One question runs through the entire story we have charted: where is vision formed? In the eye or in the brain? And if it is in the brain, in which area? When one asks oneself these questions, it is natural to imagine that man carries a homunculus inside his head, who scrutinizes the image as it arrives, placing himself first behind the lens, then contemplating the retina, and then installing himself in the brain. One has to make an effort to imagine how man functions without resorting to anthropomorphism.

The question is at what precise moment of the process does light become image? Berkeley says:

Farther, what greatly contributes to make us mistake in this matter is that when we think of the pictures in the fund of the eye, we imagine ourselves looking on the fund of another's eye, or another looking on the fund of our own eye, and beholding the pictures painted thereon. (*An Essay Towards a New Theory of Vision*)

This alternation between eye and brain continued until the microscope proved that the retina and the visual cortex are made in the same way: this was what led the way to the realization that the retina is a peripheral portion of the cerebral cortex. In short, the brain begins in the eye. (That last sentence was said by me and let's hope it's right.)

The climax of Pierantoni's book is the chapter on Camillo Golgi's discovery: I will not summarize it because that would spoil its truly remarkable effects – both artistic and dramatic.

We thus arrive at the retina as we know it today (the description is very clear but it would not have been too much to have added a drawing allowing us to see in graphic terms all the 'horizontal' and 'vertical' relationships). And the overall picture of sight which emerges from this process is such as to overturn the whole succession of previous models.

In every model Pierantoni sees 'mythical' constants, and the central thread of his book is precisely the exploding of these 'myths' on which our knowledge feeds, for they prevent us from understanding the reality of the natural processes even when we already possess all the necessary data. According to Pierantoni, the last of these mythical models is the computer.

This 'mythological' approach to the history of science and culture seems to me to be both correct and necessary: my only reservation concerns the 'polemical approach to myths' which lurks there. Knowledge always proceeds via models, analogies, symbolic images, which help us to understand up to a certain point; then they are discarded, so we turn to other models, other images, other myths. There is always a moment when a myth that truly functions deploys its full cognitive force.

The extraordinary thing is how centuries later a conception that has been rejected as mythical can reappear as fertile at a different stage of knowledge, taking on a new meaning in a new context. Would it not be right to conclude that the human mind – in science as in poetry, in philosophy as in politics and law – only functions on the basis of myths, and our only choice lies in adopting one mythical code over another? A knowledge that is outside any code does not exist: we just have to be careful to identify myths that are wearing out and becoming obstacles to knowledge, or worse still dangers to human co-existence.

Using the image of the biophysical structure of the retina in a 'mythical' way, the human mind seems to me like a tissue of 'myth-receptors' transmitting to each other their inhibitions and excitements, just like the photo-receptors which condition our sight and ensure that when we look at the stars we see them as made up of rays, whereas 'in reality' they should appear to us as dots . . .

[1982]

III Accounts of the Fantastic

The Adventures of Three Clockmakers and Three Automata

Often the commitment that men invest in activities that seem totally gratuitous, with no other aim in mind except enjoyment or the satisfaction of solving a difficult problem, turns out to be essential in an area that nobody had foreseen and has far-reaching consequences. This is true for poetry and art, just as it is for science and technology. Amusement has always been the great moving force behind culture.

The construction of automata in the eighteenth century was a precursor of the industrial revolution, which would reap the benefits of mechanical solutions that had been originally devised for complicated toys. Of course it has to be said that the construction of automata was not just a game, even though it looked like it: it was an obsession, a demiurgic dream, a philosophical challenge to put man and machine on the same level. The critical fortune of the automaton as a literary theme, from Pushkin to Poe to Villiers de l'Isle-Adam, confirms the sway of this fascination, which had both hyper-rational and unconscious elements.

All these thoughts were aroused by an unusual illustrated book published by F. M. Ricci on the 'Androids' of Neuchâtel: *Androidi. Le meraviglie meccaniche dei celebri Jaquet-Droz* (*Androids: The Mechanical Marvels of the Famous Jaquet-Droz Family*). In the eighteenth century Neuchâtel was the capital of clock-making not only in terms of artistry but also in scientific terms (see the six volumes of *Essais sur l'horlogerie* by Ferdinand Berthoud). Recently the museum of Neuchâtel has after meticulous mechanical restoration work brought back to new life three famous automata: the 'writer' or 'scribe', the 'draughtsman' and the 'female musician', constructed over 200 years ago by maestros of that tradition, the Jaquet-Droz father and son and J.-F. Leschot.

The colour plates in the volume published by Ricci document in great detail the external aspect and the internal mechanisms of the three 'Androids'; the black-and-white plates record the graphic output of the first two, and the musical scores played on the harpsichord, while the book's text tells the story of the artists and their creatures, their technical details and the recent restoration work. (Moreover, they have included in the box that contains the volumes a disk with the repertoire played by the 'musician' before and after the restoration work.)

Why on earth does such a technical and factual book provoke such a feeling of disquiet? It is true that these three 'Androids' do nothing to attenuate their doll-like appearance or to hide their machine-like nature. Perhaps one needs to go back to Baudelaire's passages on toys or Kleist's on marionettes to understand the reason for this enduring fascination. In these models the elegant and gallant eighteenth century with lace sleeves and collars and the cold, analytical eighteenth century of the diagrams in the *Encyclopédie* are both present and emphasized in an extreme form. In addition, the name 'Android' blends these hints and evokes science-fiction *avant la lettre*, as though they were a living species halfway between man and machine, or a race of possible invaders, in whom we would end up recognizing our doubles.

The 'scribe' or 'writer' is the one with the least intelligent face but the most complicated mechanism: his wrist moves in three directions, the quill pen traces the letters with calligraphic lines and loops, dips into the ink-well and goes to the next line like a typewriter; a device stops it when it writes a full-stop. A series of cams allows it to write the letters of the alphabet, small and capital, and to compose the sentences written into the program.

The performances carried out by the 'draughtsman' are on the surface more showy, but his mechanism is much less complicated than that of the 'writer'. His repertoire consists of four drawings, closely tied to the age in which he was made: one of them is a little dog, and another is a profile of Louis XV. The story has it that on the occasion of a 'performance' in the presence of Louis XVI and Marie Antoinette the nervous operator announced that a drawing of the late king would appear, but he did the wrong thing when starting the mechanism: the automaton's pencil slowly drew the little dog, 'the which matter spread a certain disquiet'.

While the two graphic geniuses have the faces of big infantile dolls, the woman-doll playing the harpsichord has such a mysterious expression and look that one can imagine perverse stories of people falling in love with her, as in works by Tommaso Landolfi or Felisberto Hernández. The author of the commentary in the book explains that she is 'the only doll in the world who breathes, thus sharing our life, apparently drawing the source of her existence from the same air that we depend on', and wonders whether she was not meant to be 'offering herself through her delicate music to a lover fantasizing about unreal delights, or reviving in Pierre Jaquet-Droz the immortal memory of his young bride whom he had lost for ever . . .'

The story of Pierre Jaquet-Droz (1722–90) makes a fine, typically eighteenth-century biography. In order to dedicate himself to clockmaking he abandoned his theological studies. He perfected his art with frequent stays in Paris (where already in the previous generation some maestros from Neuchâtel had established themselves as Court clock-makers), and found a base at the University of Basle working with Johann Bernoulli and other members of that famous family of mathematicians.

Jaquet-Droz's fame soon spread from the Jura mountains to the rest of Europe. Neuchâtel in those days, although part of the Swiss Federation, was a princedom subject to the King of Prussia, and its closest links were relations with foreign Courts. With a cart full of his *pendules* Jaquet-Droz went all the way to Madrid, and obtained official recognition of his craftsmanship from the Court of Spain.

Back in his native land, he set up a laboratory at La-Chaux-de-Fonds with his son Henri-Louis (1752–91) and his adoptive son Jean-Frédéric Leschot (1746–1824). He was by now the head of an established firm, and it was at this point, at the height of his fortune, that he decided to build the 'Androids'. Who provided the decisive impulse? Was it the Bernoulli family? Was it a local doctor whom chronicles of the time describe as part-inventor, part-naturalist, part-magician? Was it Leschot, whose portrait shows us the face of a wise gnome (whereas the portraits of the Jaquet-Droz, father and son, are rather inexpressive)?

Whoever it was, the fact is that after 1773–4, the date of the construction of the three automata, the life of the three clock-makers changed: they lived mostly for their creatures, showing them to illustrious visitors and taking them on tour throughout the various European capitals. But at the same time their business expanded: they founded a branch in London to export precious clocks, carillons, singing birds and other mechanical wonders to China and India.

However, some confusion started to arise: when people said 'the Droz' were they talking about the three clock-makers or the three automata? 'The three Droz' by now meant the latter: that is the way we see them in a print of the period; the three mechanical dolls took the names and surnames of members of the family. I do not know the precise date of the print: was it before or after the fall of the Bastille? One might almost say that the automata rebelled and claimed their independence by usurping the identity of their inventors.

Was this the reason that the great Jaquet-Droz firm ran into a slump and rapidly went bankrupt? Certainly the French Revolution hit the luxury goods market hard and the Napoleonic wars ruined exports; but the crisis had apparently happened earlier, a crisis which affected the whole Swiss watch-making industry.

It is also a fact that in 1789 the 'Androids' no longer appear in the firm's inventory. They went from hand to hand, constantly being exhibited to the public as a spectacle and attraction. (Or was it the automata themselves that, after proclaiming the 'rights of the automaton', then moved freely throughout Europe?) In their tours they ended up in Saragossa, which was being besieged by Napoleonic troops, and they were then captured and carried off to France with other war booty. They then resumed their wanderings in international exhibitions, which continued throughout the nineteenth century.

The story contains a unique display of loyalty: throughout the whole of the nineteenth century the citizens of Neuchâtel never forgot about the existence of their three children who had become lost somewhere in the world. Every so often local papers would run appeals to track them down and recover them. This duly happened in 1905 thanks to a public subscription. (Or was it the automata themselves who wanted to return to their native land? They had started their wanderings in the footprints of the great adventurers of their century, indefatigable optimists like Cagliostro, Casanova, Candide. But at the dawn of the new century they realized in time that the world was about to become impossible for those whose movements were controlled by vital mechanisms that were so simple and transparent. It was just as well they remembered that they were Swiss citizens before it was too late.) The following sentence was inserted into the programme of 'the writer', and he still copies it out in his eighteenth-century handwriting: 'We shall never leave our country again.'

[1980]

Fairy Geography

Their first attribute is lightness. They are small in stature, with bodies 'somewhat of the Nature of a condensed Cloud' or 'of congealled Air', in short, of matter so thin and slender that for nourishment all they need is any liquid that can penetrate their pores as it would into a sponge, or bits of grain that they fight over with crows and mice. They live beneath the earth, in little mounds perforated by tunnels and fissures, but sometimes they are borne aloft, flying in mid-air. Their appearance and perhaps their very presence is discontinuous: only those gifted with second sight can see them, and always for brief moments as they appear and disappear. Their underground domains are illuminated by perpetual lamps which shine without any combustible matter; some say their very persons emit a greenish light. They have lives that last much longer than human lives, but they too are mortal: at a certain point, without any illness or suffering, they become rarefied and disappear . . .

They have been known to work, if it is true that near their dwellings 'they are some times heard to bake Bread' and 'strike Hammers'. Their women weave and sew 'strange Cobwebs', according to some, 'impalpable Rainbows', according to others, or clothes similar to our own, according to others still. But also in our own kitchens, sometimes while we are sleeping, it is they who dutifully tidy up the dishes for us and put everything back in place. Their relationships with human beings consist in these little duties but also in spiteful acts and petty theft, or in throwing stones, some of them quite big, which, however, do not hurt. More seriously they kidnap babies, or wet-nurses (they are greedy for milk) who stay a certain time with them underground while up here their person has been replaced by a double or a ghostly presence.

They also have sexual relations with humans, especially with women, but more at the level of a fleeting, lascivious game, like in dreams, without passion or drama. They are not without wars or cruelty, but they keep it all to themselves and let little of it out. They use the human language of the place where they live, but 'they speak but little, and that by way of whistling'. 'They are said to have many pleasant toyish Books; but the operation of these Pieces only appears in some Paroxisms of antic corybantic Jolity.' They have moments of jubilation and edginess, but their most frequent mood is melancholic, perhaps due to their in-between nature.

The people I am talking about are the 'little People' of the Siths, the subject of a book that Adelphi has published (the Italian translation of Robert Kirk's *The Secret Commonwealth of Elves, Fauns and Fairies*, edited by Mario M. Rossi, whose essay 'The Fairy Chaplain', closes the book). 'Siths' is the name that was given in Scotland to what are called 'fairies' in England (there is no exact equivalent in Italian since *le fate* for us can only be feminine whereas a 'fairy' can be either male or female). In the Anglo-Germanic world they are known as 'elfs' or with some particular differences 'kobolds' or 'goblins', and they include all the varieties of dwarves and gnomes (often connected with mines and hidden treasures), including Tolkien's 'hobbits'.

The supernatural of the Celtic peoples is swarming and intricate and multiform, difficult to put into a proper order. Or perhaps it is we Latins who see the Mediterranean world of fauns and nymphs and dryads and hamadryads as more orderly, solely because our luxuriant local mythologies have been put through the sieve of hierarchical organization and harmonization typical of Graeco-Latin culture. The power of poetic transfiguration in the Nordic imagination has given us Titania and Oberon and Puck, as well as Spenser's *The Faerie Queen*. But even when transmitted through the words of the poets the kingdom of Celtic fairies communicates the pristine power of a world that is irreducibly 'other', that literature cannot tame entirely.

Also in the Celtic parts of France (especially Brittany and Normandy) the 'little people' have ancient roots, and in literature they have left traces in the fantastic tales by Nodier and in a novel by Barbey d'Aurevilly, *L'Ensorcelée* (The Bewitched), where the emerging of magic-telluric apparitions is conveyed in the most disturbing way. But it is in the green fields of Ireland and the Scottish moors that this elusive race reached its highest population density. Walter Scott for Scotland (in his *Letters on Demonology and Witchcraft*) and W. B. Yeats for Ireland (in *Irish Fairy and Folktales*) attempted, if not a census, at least a classification of species and families: these two geniuses brought a systematic mentality to the cult of these traditions.

The case of Robert Kirk is different. At the end of the seventeenth century he was minister of the Presbyterian church in Aberfoyle, a village at the entrance to the Highlands. Scotland had recently been subdued by the English Crown and devastated by civil wars and the wars of religion, and such impoverished communities lived through a permanent survival crisis, as well as a crisis of cultural and religious identity. We are talking of a time and place where the survival of the old beliefs was strongly felt, the topography of the place itself was steeped in the presence of fairies, and 'second sight' was a common experience. But these were also times and places where the battles between Anglicanism and Presbyterianism had implications that were very much political as well as theological.

The seventeenth century was the century of witch trials and inquisitors (both Catholic and Protestant) who saw in the various forms of the pre-Christian supernatural that still survived nothing other than the uniform presence of Satan, to be eradicated by burning. The Reverend Kirk had a strong and profound innate purity that enabled him to recognize clearly his neighbour's innocence. He knew that his parishioners who believed in fairies and could see them were not witches and wizards; he was fond of his poor Scottish flock and knew their hallucinations and the precariousness of their existence; he was fond of the fairies, who were a poor people as well, and were perhaps on the verge of disappearing, without a physical or metaphysical place to call their own. He too certainly believed in fairies, and probably saw them, even though he restricted himself to quoting the evidence of other people.

With the courage of innocence he wrote a treatise on the fairies' kingdom, entitled *The Secret Commonwealth*, in order to say all he knew about them, which was not much, and above all to allay any suspicion of collusion with the devil in the little underground fairies and in those who saw them. (Here, in addition to the problem of the fairies' existence there was also the question of second sight, telepathy, premonitions, phenomena which are not necessarily – indeed are rarely – connected with mediation by supernatural beings.) The quotations from Sacred Scripture with which Kirk supports his argument are inexact and never entirely relevant, but his intention is clear. He wanted to establish that 'the little people' had nothing to do with Christianity but neither did they have anything to do with the devil: their legal position was the same as Adam's before the Fall, so they would be neither saved nor damned; it was a neutral, unjudgemental limbo that surrounded their sins, which were always minor and childlike, and their melancholy.

The book that Adelphi has recently published contains Kirk's treatise, discovered and translated by Mario Manlio Rossi, plus a substantial essay by the latter: with both erudition and passion he situates the work in the culture of its time, and explains in exhaustive terms how Kirk genuinely believed in the fairies and how there was nothing strange in that. There are thus three sources of interest in the book: the fairies themselves, the personality of this 'chaplain of the fairies', and the personality of the man who discovered and explained Kirk.

Mario Manlio Rossi (1895–1971), an Italian scholar of English literature who lived in Edinburgh for many years, was a reserved academic who always went against the grain. I know little about him but I will always be in his debt, for it was through a book of his that, when I was young, I understood the greatness of Swift. Rossi argues effectively here that trials for witchcraft were not in fact a medieval throw-back but instead a typical product of modern culture. His essay is fascinating for the richness of detail that it evokes and documents in this depiction of the history of culture, but it is also readable for the polemical mood – often a bad mood – that emerges on every page, evidence of a surly temperament in which learned precision blends with *parti pris*. His anger has a range of targets – both Presbyterian and Anglican intolerance, witch-hunts and the opinions of all historians who have dealt with them, fairy-tales for children that censure the sexual element which is always present in popular stories – but he also attacks Empiricism, Idealism, Occultism, Folklore and above all Science, which is his *bête noire*. He only spares (and here I have no doubts in agreeing with him) poetry, where 'man in flesh and blood and the fairies have the same epistemological position, the same reality'.

As I was reading the book the name of Kirk's village, Aberfoyle, continued to ring a bell in my head. Why did it sound familiar? Of course, that was the setting of my favourite Jules Verne novel, *Black Diamonds or The Child of the Cavern*. It was a story that took place entirely underground, in an old, abandoned coal-mine, the hiding-place for creatures that seem to have come straight out of the Reverend Kirk's book: a fairy-girl who has never seen the light of day, a venerable old man who resembles a ghost, a horrible bird from the abyss . . . Suddenly the Celtic visionary world infiltrates the apologia for science written by the positivist Verne, to show that, despite Mario Manlio Rossi, the same mythological lymph flows and mixes in the unfathomable tangle of ideologies that are apparently opposed to each other . . . All of this proves that the fairies know even more roads, in heaven and earth, than are dreamed of in any of our philosophies . . .

[1980]

The Archipelago of Imaginary Places

On Frivola, an island in the Pacific, life is easy and frustrating. The trees are as elastic as rubber and their branches bend down to offer fruits that melt in the mouth like froth. The inhabitants rear fragile and useless horses which collapse under the slightest weight. To plough the fields, all that is needed is for the women to play on a whistle and furrows open up in the thin dust, while in order to sow men just scatter seeds to the wind. In the forests the wild beasts have soft tusks and claws and their roar is like a rustle of silk. The local currency is the *agatina*, which is not very prized on the currency market.

The Diamond Islands have the property of swallowing up imprudent travellers, who are captured by their carnivorous diamonds. In order to get hold of the jewels, crafty merchants scatter bloody pieces of pork over them, which the diamonds immediately start to suck on; towards evening the vultures descend, snatch the meat in their claws and fly off with it to their nests, along with the jewels stuck to the pork. The merchants climb up to the nests, frighten off the raptors, separate the diamonds from the meat, and then sell them to unwary jewellers. That is how a ring devours a finger, or a necklace a neck.

Capillaria, a land beneath the sea, is inhabited exclusively by self-reproducing women called Ohias: they are beautiful and majestic, two metres tall, with features like angels, soft bodies and long blonde hair framing their faces. The Ohias' skin feels like silk, and is translucent, like alabaster: through their transparent skin you can see the bones of their skeleton, their blue lungs, their pink heart, the calm pulsing of their veins. Men are unknown there, or rather they survive as external parasites called Bullpops, formed of a cylindrical body about fifteen centimetres long, a bald, bumpy head, a human face and wiry arms and hands, but they have legs endowed with huge big toes, fins and wings. The defenceless Bullpops swim vertically like sea-horses, and the Ohias feed on them since they are greedy for their marrow, to which they attribute amongst other things properties that somehow stimulate reproduction.

On the island of Odes the roads are living creatures and they move freely of their own accord. To travel across the island visitors just have to take up their position on a road, after finding out where it is going, and let themselves be carried along. The most famous roads in the world come to Odes as tourists for a holiday.

London-on-Thames, which is not to be confused with its more famous namesake, is a city dug out of the top of a rock, inhabited by a tribe of gorillas whose chief believes he is the reincarnation of Henry VIII and he has five wives called Catherine of Aragon, Ann Boleyn and so on. The sixth wife is a white woman, captured by the gorillas, who stays in this role until she is substituted by another female captive.

On the island of Dionysus there is a vineyard growing where the vines are women from the waist upwards; vine-leaves and clusters of grapes dangle from their fingers, and their hair is made of tendrils. Heaven help the traveller who allows himself to be embraced by these creatures: he immediately gets drunk, forgets his homeland, family and honour, puts down roots and becomes a vine as well.

Malacovia is a fortified city made entirely out of iron, and built on the Danube Delta: it is in the shape of an egg, chock-full of Tartar cyclists who, as they pedal, make the iron egg go down, so it is concealed in the Delta marshes, and then back up again. The city lives, waiting for the moment when the hordes of cycling Tartars will be unleashed to invade the empire of the Czars.

The sources of these geographical descriptions are respectively:

Abbé François Coyer, *The Frivolous Island* (London, 1750); *The Thousand and One Nights*; Frigyes Karinthy, *Capillaria* (Budapest, 1921); Rabelais, *The Fifth Book of Pantagruel*; Edgar Rice Burroughs, *Tarzan and the Lion-Man*; Lucian of Samosata's *True Story*; Amedeo Tosetti, *Pedali sul Mar Nero* (Pedals by the Black Sea) (Milan, 1884).

That, at least, is how they are cited (I take no responsibility for their veracity) in the book from which I drew this information: *The Dictionary of Imaginary Places*, by Alberto Manguel and Gianni Guadalupi (Toronto: Lester and Orpen Dennys, 1980). This is an enormous volume with the layout of a geographical dictionary and entries in alphabetical order (from Abaton, a city that has a variety of geographical locations, to Zuy, the Elves' shopping centre), and it comes complete with maps and engravings like those of an old-fashioned encyclopedia.

A book published in Canada and the product of a collaboration between an Argentine and an Italian has all the credentials for epitomizing geographical confusion. In the Library of the Superfluous, which I would like all our bookshelves to find a space for, it seems to me that a Dictionary of Imaginary Places would be an indispensable reference work.

Every city or island or region has an entry as in an encyclopedia, and every entry begins with information on its geographical position, population and any economic resources, as well as its climate, fauna and flora. The rule behind the Dictionary is to present every place as though it really existed. This information is derived from the sources, which are given at the end of every entry: thus for Atlantis are listed Plato's *Critias* and *Timaeus*, Pierre Benoît's novel and also a lesser known work by Conan Doyle.

Another rule that the authors obey is to exclude imaginary toponyms used by novelists to represent real or at least probable places: so Proust's Balbec is not there, nor Faulkner's Yoknapatawpha. And given that the geography concerns the present and the past but not the future, the whole of futuristic science-fiction, whether extraterrestrial or political or social fantasy, is also excluded.

This is not a book that hooks you immediately. On the contrary, the first impression as you thumb through it is that imaginary geography is much less attractive than the geography of real places: a methodical dullness hangs over utopian cities, from Francis Bacon's Bensalem to Cabet's Icaria, as well as over countless eighteenth-century satirical-philosophical voyages, not to mention the edifying religious-allegorical stages along Bunyan's *Pilgrim's Progress*. And a sense of satiety, not to say lack of oxygen, accompanies the packed topographies in *The Wizard of Oz*, Tolkien or C. S. Lewis.

However, as one works one's way through the single entries one soon comes across worlds that are governed by a more evocative fantasy logic, and I have tried to provide some examples of these above; I have not quoted (because it is already well known in Italy thanks to Masolino d'Amico and Giorgio Manganelli) what remains the most elegant and ingenious invention: Abbot's geometrical Flatland.

It is above all minor literary fiction that reveals endless resources for creating these poetic myths; whole atlases of visionary countries flow from the pen of talented professionals in entertainment literature. The most quoted author is Edgar Rice Burroughs, not only for his cycle of Tarzan books but for a large number of works describing fantasy lands. Taken from novels that were considered merely as page-turners and whose authors are not recorded in literary histories, many such states went on to become myths of the cinema such as the Shangri-La of *Lost Horizon*, the Ruritania of *The Prisoner of Zenda* and the Island of Count Zaroff in *The Most Dangerous Game*. The Dictionary also includes countries that were created directly for the screen, such as the Marx Brothers' Freedonia in *Duck Soup*, and Pepperland in the Beatles' *Yellow Submarine*; however, I do not see the cities from René Clair's films of political satire.

Italian literature is well represented, from Boiardo's Albraca to Zavattinia in *Totò il buono*, even though it is not the richest in this field: still the Bastiani Fortress in Buzzati's *The Tartar Steppe* is there, as is Gadda's Maradagal and Pinocchio's Toyland. Amongst the curiosities worth mentioning I will point out two tunnels: one that leads from Greece to Naples, for the exclusive use of unhappy lovers, which is explored in Sannazaro's *Arcadia*; and the other that links the Adriatic (through the valley of the river Brenta) to the Tyrrhenian Sea (leading to the Gulf of La Spezia), constructed in the fourteenth century by the Genoese in order to invade the Republic of Venice. The latter was tracked down and explored in Salgari's novel, *I naviganti della Meloria* (The Sailors of the Meloria) (1903): in the novel the sailors actually found in the tunnel phosphorescent fauna consisting of jellyfish and giant molluscs.

[1981]

Stamps from States of Mind

Throughout his life Donald Evans made stamps. Imaginary stamps of imaginary countries, drawn with pencils or coloured inks and painted in watercolours, but scrupulously faithful to everything one would expect from a stamp, to the point where they seemed, at first sight, genuine. He would invent the name of a country, the name of a currency, a range of imaginary sights, and would start to insert minute details into tiny quadrangles or squares (sometimes triangles), all of them framed with a white, perforated border. He would produce complete series, each of which had its year of printing and the style of the period and contained stamps of every value in their delicate little shades, selected from the usual range of colours you find in postage stamps.

Nothing to do with science-fiction, not utopian or weird: the states of his imaginary atlas resemble states that exist in reality, and are rendered that bit more familiar and accessible by being associated with a small number of reassuring emblems. He would also invent a name for the capital and would superimpose a circular franking mark on the stamps, so that they looked even more persuasively lifelike. At times his work would also include an envelope completely covered in stamps and franking marks, with the address written in mock handwriting, and names of people and places also invented but always just about credible.

The fascination of stamps always starts in childhood: it is aroused both by a passion for the exotic and by an obsession with systematic classification. It was as a child that Donald Evans, an American from New Jersey, began not only to collect stamps but also to invent new ones, which meant inventing a geography and history parallel to the geography and history of the world that other people recognize. As he grew up, Evans did not completely abandon this childhood passion, even though, while continuing to paint in any spare time he had from his course on architecture, he did hide it as though ashamed of it. This was in New York at the end of the 1950s, a period when Abstract Expressionism totally dominated. But a short time later the arrival of Pop Art persuaded Evans that his earliest figurative predilections were in tune with the most up-to-date artistic tastes. The road to success as a painter opened up before him: but instead the only thing that interested him was finding the peace to live doing what he liked best. In the 1970s he did nothing but paint stamps, about 4,000 of them, distributed across forty-two imaginary countries, with an exhibition every year, but he stayed in New York as little as possible. He lived almost permanently in Europe, especially in Holland, up until the fire which cost him his life, in Amsterdam, at just thirty-one years of age. The book that introduced me to him is proof that a circle of friends and connoisseurs venerates him and his work as though he were a saint (*The World of Donald Evans*, text by Willy Eisenhart, New York).

Willy Eisenhart reconstructs the short life of Donald Evans (1945–77) in minute detail and comments on his oeuvre, all this by way of an introduction to the eighty-five colour plates which are organized as in a collector's album: the imaginary countries are in alphabetical order. This collection of stamps is also at the same time a collection of hens, windmills, dirigibles, chairs, palms, butterflies and all sorts of exemplars of the various countries' flora and fauna (actually 'Flora and Fauna' is the name of a federal country which is situated goodness knows where in Evans's geography, certainly in Nordic climes). The fact is that Evans adores classifications, nomenclature, catalogues and pattern books: and what better form could this serial passion of his take than a whole series of stamps? 'Catalogue of the World' is the title that he proposed for his entire oeuvre.

Other pages depict sheets of stamps which are all identical and not yet separated along the perforated lines. Others still show stamp collections that try to reconstitute this kind of original sheet by putting in rows stamps that are all the same but are differentiated by the dark shadow of the frank mark and the irregularities of their outlines. (Evans took particular care in imitating the perforations, or the absence of perforation in series that are meant to be older and that precede the invention of the perforating machine.) There are also more abstract combinations, like the domino pieces in the extremely elegant stamps of 'Etat Domino', or the Scottish tartans of 'Antiqua', which were painted in honour of a female friend whose family originated from Scotland.

Eisenhart sees the origin of this philatelic obsession in Evans's introverted character. I would say that what inspired him was the urge to keep a diary of states of mind, feelings, positive experiences, values that were summed up in emblematic objects; but the nostalgic vision of the stamp album allowed him to cultivate an interiority that had at the same time become objectivized and dominated by his consciousness. What prevails is the order in this serial arrangement, the irony of his invention and attribution of names, and also the restrained melancholy of hazy landscapes, repeated in all the different colours.

Creating stamps was for Donald Evans above all a way of appropriating the countries he had visited, the places where people lived: his adopted homeland, Holland, inspired him to create the stamps of 'Achterdijk' ('Behind the Dyke', from his first Dutch address), and of 'Nadorp' ('After the Village', from a friend's address), where he expressed his love for the flat landscapes, windmills of various kinds and also the Dutch language. The stamps of 'Barcentrum', from the name of a bar Evans frequented in Amsterdam, have more lively colours: this is a beautiful series which is also a list of drinks in order of price, in glasses that are all different. Gradually we realize that many of these names of states are not invented at all, but are the names of modest or tiny places which Evans had been to and to which he attributed the prerogatives that belong to sovereign states. Thus after a summer on the Costa Brava he designed the stamps of Cadaqués, with a cheerful series of vegetables.

Other names belong to a geography of feelings: 'Lichaam' and 'Geest' ('Body' and 'Soul', in Dutch) are two twin kingdoms in the far North which have the same currency (the 'ijs', in other words ice) and stamps (with seals and narwhals). Two African islands are called 'Amis et Amants' and form one of the states that emerged from the decolonialization of an ancient French protectorate, the 'Royaume de Caluda'. Initially the new states still used the dreary stamps of the old colony with the changed name overstamped on them; then the 'Postes des Iles Amis et Amants' issued a new series with views of places called 'Coup de Foudre', 'Premiers Amours' and 'La Passade'.

But it was above all through food that Evans established his relationship with countries, catching their most typical flavours and aromas during his travels. After a trip to Italy he invented a new country, 'Mangiare', whose currency was calculated in grams and whose very sophisticated stamps form a museum of vegetables, fruit and herbs: from peas, capers, pine-nuts, olives (small dot-like images which stand out against a bare background and are elegantly framed) to courgette-flowers, rosemary, celery and broccoli. The state of 'Mangiare' dedicated a special issue to the recipe for pesto alla genovese, with its basic ingredients (basil, pine-nuts, pecorino cheese, garlic). Another series (dated 1927) exalted the cucumber, which was portrayed in the shape of a dirigible. During the Second World War the state of 'Mangiare' was invaded by the army from Antipasto: an overstamp indicates the stamps from the occupied zone. In the postwar period a region of 'Mangiare' called 'Pasta' became independent: the 'Poste Paste' issued a series which is a splendid showcase of varieties of pasta.

Even the homesickness experienced by the American in Europe is focused on visions of food such as fruit. The evocative plates devoted to a country called 'My Bonnie' (as in the song 'My Bonnie Lies over the Ocean') are dotted with cherries, all apparently identical, but each with a different shade of red and a different name, taken from catalogues produced by fruit farms.

All in all, this supposed introvert was a man who was not at all turned in on himself but projected outwards, towards the things of the world, chosen and recognized and named one by one with great delicacy and loving precision. Perhaps what interested him most in stamps was precisely their celebratory function: he wanted to oppose the carefully programmed, bureaucratic, official celebrations of all the postal ministries in the world with a ritual of private celebrations, commemorations of minimal encounters, consecrations of things that are unique and irreplaceable: basil, a butterfly, an olive. Without the illusion of stealing them from the flow of time which rapidly transforms stamp series into traces of the past.

[1981]

The Encyclopedia of a Visionary

In the beginning was language. In the universe that Luigi Serafini inhabits and describes, I believe that the images were preceded by the written word, by those minute, agile and (we have to admit it) very clear italics of his which we always feel we are just an inch away from being able to read and yet which elude us in every word and letter. The anguish that this Other Universe conveys to us does not stem so much from its difference to our world as from its similarity: similarly the writing could easily have been developed in a linguistic area that is foreign to us but not unknowable.

On reflection, it occurs to me that the peculiarity of Serafini's language cannot be solely in its alphabet but also in its syntax: the things belonging to the universe that this language evokes, as we see them illustrated in the plates of his encyclopedia (*Codex Seraphinianus*, published by Franco Maria Ricci), are almost always recognizable, but it is the connection between them that seems to be turned inside out, with unexpected combinations and relations. (If I said 'almost always' that was because there are also some unrecognizable forms, and these have a very important function, as I shall try to explain later on.) The crucial point is this: if Serafinian writing has the power to evoke a world where the syntax of things has been distorted, it must contain, hidden beneath the mystery of its indecipherable surface, a deeper mystery still regarding the internal logic of language and thought. The images of things that exist coil and link together; the havoc this wreaks on visual attributes generates monsters; Serafini's universe is inhabited by freaks. But even in the world of monsters there is a logic whose outlines we seem to see emerging and vanishing, like the meanings of those words of his that are diligently copied out by his pen-nib.

Just like Ovid in the *Metamorphoses*, Serafini believes in the contiguity and permeability of every territory of existence. The anatomical and the mechanical swap morphologies: human arms, instead of finishing in a hand, finish in a hammer or a pincer; legs are supported not by feet but by wheels. The human and the vegetable complete each other, as in the illustration of the cultivation of the human body: a wood on its head, climbing plants up the legs, lawns on the palm of a hand, carnations flowering out of its ears. The vegetal world also mixes with the mercantile one (there are plants with a trunk-cum-wrapped-sweet, others with ears in the shape of pencils, others with leaves as scissors, or with fruit like matches), the zoological mingles with the mineral (dogs and horses that are half-petrified), as does cement with geology, the heraldic with the technological, the savage with the metropolitan, the written with the living. Just as certain animals take on the form of other species living in the same habitat, so living beings are infected by the forms of the objects that surround them.

The shift from one form to another is followed stage by stage in the human couple making love, who gradually metamorphose into an alligator. This is one of Serafini's most ingenious visual inventions, alongside which I would put, in my ideal selection, those fish which are just surfacing from the water and seem to become the huge eyes of a goddess of the silver screen; and also the plants that grow in the shape of a chair, so that all you have to do is to cut them and whittle them down in order to have a ready-made wicker seat; and I would add yet one more thing: all the figures in which the motif of the rainbow appears.

I would say that there are three images that most inspire Serafini's visionary raptus: the skeleton, the egg and the rainbow. One would think that the skeleton is the only nucleus of reality that stays as it is in this world of interchangeable forms: we see skeletons waiting to put on their covering of skin and flesh (which hang limply from hooks, like empty suits of clothes) and after the act of getting dressed look at the mirror, perplexed. Another illustration conjures up a whole city of skeletons, with television aerials made of bones, and a skeleton-waiter serving a bone on a plate.

The egg is the fundamental element that appears in all its forms, with or without its shell. Shell-less eggs fall from a tube on to a lawn, which they immediately cross, slithering along like organisms endowed with perfect autonomy of locomotion, only to then climb up a tree and fall again, this time taking on the characteristic shape of fried eggs.

As for the rainbow, it occupies a place of central importance in Serafini's cosmology. As a solid bridge, it can support an entire city; but it has to be said that this is a city that changes colour and consistency, just like its support. It is from the rainbow that certain little animals emerge, multi-coloured and two-dimensional, with irregular shapes never seen before: they emerge from circular holes in the iridescent tube, and could be the real vital principle of this universe, corpuscles that generate the unstoppable general metamorphosis. In other illustrations we see that what is spraying out the rainbows into the sky is a kind of helicopter which can draw them in their classic semi-circular form but also in the shape of a knot, a zig-zag, a spiral, or as a steady series of drips. From this helicopter's fuselage, in the form of a cloud, hang so many of those polychrome corpuscles, attached to wires. Are these corpuscles the mechanical equivalent of the iridescent dust-cloud which is suspended in the air? Or are they hooks placed there to catch colours?

These corpuscles are the only indefinable shapes in Serafini's visual cosmos, as I mentioned earlier. Beings of similar form appear like luminous little bodies (are they photons?) in a swarm flying out from a headlight, or like micro-organisms carefully catalogued at the opening of the botanical and zoological section of this encyclopedia. Perhaps they have the same consistency as graphic signs: they constitute yet another alphabet, more mysterious and archaic. (Similar shapes, in fact, appear sculpted on a kind of Rosetta Stone, alongside their 'translation'.) Maybe everything that Serafini shows is a kind of writing: only its code changes.

In Serafini's writing-universe almost identical roots are catalogued with different names because every tiny little root is a differential sign. The plants twist their tender stalks like lines traced by the pen, they penetrate the earth from which they have just sprouted, only to bring forth subterranean blooms or to resurface once more.

These vegetal forms continue the classification of imaginary plants that was begun by the genteel Nonsense Botany of Edward Lear and continued by Leo Lionni's astral Parallel Botany. In Serafini's nursery there are cloud-leaves that water the flowers, and webleaves that capture insects. Trees uproot themselves spontaneously and walk; they go to the seashore from where they set sail, their roots whirring like propellers on a motor-boat.

Serafini's zoology is always disturbing, monstrous, the stuff of nightmares, a zoology whose evolutionary laws are the metaphor (a sausage-snake, a shoelace viper on a tennis shoe), metonymy (a bird which is a single wing ending in a bird's head), and the condensation of images (a pigeon that is still an egg).

After the zoological monsters come the anthropomorphic ones, perhaps failed efforts on the road to humanization. That man became man starting from the feet upwards was explained by the great anthropologist Leroi-Gourhan. In Serafini's illustrations we see a series of human legs that try to find completion not in a torso but in an object such as a ball or an umbrella, or just in a luminosity like that of a star shining and going out. It is a crowd of beings of this latter kind that we see standing on boats drifting down a river, passing under the arches of a bridge, in one of the most mysterious images of the volume.

Physics, chemistry and mineralogy inspire Serafini's most relaxing pictures: most relaxing because most abstract. But the nightmare starts up again with mechanics and technology, where the tendency of machines to morph into monsters is no less disturbing than the human tendency to do the same thing. (Here we are reminded of Bruno Munari and a whole group of inventors of crazy machines.)

If we move to the human sciences (including ethnography, history, gastronomy, games, sport, clothing, linguistics, urban studies) we have to bear in mind that it is difficult to separate man as subject from objects which are now soldered on to him in an anatomical continuity. There is also a perfect machine which satisfies all man's needs, and at his death becomes a coffin. Ethnography is no less horrific than other fields: amidst the various kinds of savages, catalogued with their characteristic costumes and weapons and dwellings, there is the rubbish-man and the rat-man, but the most striking of all is the man in the street or rather the street-man, with a suit of asphalt decorated with the white lines of road markings.

There is an anguish in Serafini's imagination which perhaps reaches its peak in gastronomy. And yet here too we discover his peculiar kind of happiness, which comes out above all in technological inventions: a plate with teeth that chews food so that it can be sucked up in a straw; a contraption for supplying fish as though they were running water, through tubes and taps, so as to maintain a constant supply of fresh fish at home.

It seems to me that the 'gay science' for Serafini is linguistics. (Especially as regards the written word; whereas the spoken word still conjures up some anguish for him, as we see it dripping from lips like a blackish mush, or being extracted with fishing rods from an open mouth.) The written word is alive too (you just have to prick it with a hatpin to see it start to bleed), but it enjoys its autonomy and physicality, it can become three-dimensional, polychrome, can rise up from the page hanging on to balloons, or drop on to it in parachutes. There are words that, in order to stay attached to the page, have to be sewn on to it, the thread passing through the loops in those letters that have spaces in them. And if you look at the writing with a lens, the thin sliver of ink turns out to be permeated with a thick flow of meaning: like a motorway, like a swarming crowd, like a river brimming with fish.

In the end (and this is the last illustration of the *Codex*) the fate of all writing is to collapse into dust, and also all that remains of the writing hand is its skeleton. Lines and words detach themselves from the page, start to crumble, and from the little piles of dust suddenly the little rainbow-coloured beings spring out and start to jump. The vital principle of all the metamorphoses and all the alphabets starts its cycle again.

[1982]

IV The Shape of Time

Japan

The Old Woman in the Purple Kimono

I am waiting for the train from Tokyo to Kyoto. On the station platform in Tokyo there are signs showing the exact spot where the
doors of each carriage will be when the train stops. The seats have
all been pre-booked and even before the train arrives the passengers
are all at their right places, queuing between the white lines that hive
off so many little queues at right angles to the tracks.

Agitation, confusion, irritation all seem to be absent from Japanese stations. Those who are departing arrange themselves as
though on a chessboard where all the moves have been dictated in
advance. And those arriving are conveyed in flows of compact, solid,
continuous humanity which pour down escalators that leave no
space for disorder: every day millions of people go by train from
home to work in the endless space that is Tokyo.

Amongst the departing passengers in their queues I notice an
elderly lady in a rich, pale-purple kimono, surrounded by younger
members of her family, male and female, all looking respectful and
anxious for her. The farewells of families at stations are like scenes
from another age, especially in an age like ours, which is full of constant comings and goings, so much so that commuter movements
constitute the norm. In airports the ritual of farewells and greetings
which define the journey as something exceptional can still provide
material for a study of emotional behaviour in the various countries
of the world, but railway stations are becoming more and more the
realm of lonely crowds where nobody accompanies anyone else. All
the more so for a train like this, which is only going as far as Kyoto,
three hours away.

New to the country, I am still at the stage where everything I see has a value precisely because I don't know what value to give it. All it would need would be for me to stay a while in Japan and undoubtedly I too would find it normal that people should greet each other with a series of deep bows, even at the station; that many women, especially the older ones, wear a kimono with a lavish bow at the back which forms a slight hump under their overcoat and that they walk with tiny trotting steps of their white-socked feet. When everything finds an order and a place in my mind then I will start not to find anything worthy of note, not to see any more what I am seeing. Because seeing means perceiving differences, and as soon as differences all become uniform in what is predictable and everyday, our gaze simply runs over a smooth surface devoid of anything to catch hold of. Travelling does not help us much in understanding (I've known this for a while; I did not need to come to the Far East to convince myself that this was true) but it does serve to reactivate for a second the use of our eyes, the visual reading of the world.

The old lady has taken her place in the carriage alongside a young girl about twenty years old, and now they are exchanging elaborate bows with those they have left on the station platform. The girl is pretty, smiling, wearing over her kimono a kind of bright tunic, made of light fabric, which could be a housecoat, an apron. Whatever it is, she conveys a homely air, perhaps solely because of the way she is making the older woman's seat into a cosy corner, taking out of her luggage baskets, flasks, books, magazines, sweets – all the things that make a journey comfortable. This girl has nothing Western about her; she is an apparition from another age (who knows which?), with her hair-style and her laughing, fresh, gentle look. On the other hand, as far as the older woman is concerned, those few Western or rather American elements – glasses with a silver frame, the bluish perm straight from the hairdresser's – which sit on top of the traditional costume provide a clear snapshot of modern Japan.

The carriage has many spare seats, and the girl, instead of sitting next to the woman, has sat down in the row of seats in front of her and is popping up over the back of the seat to serve her food: a sandwich in a small straw basket. (Western food in a traditional Japanese container, this time – the opposite of what one usually sees in the frequent snacks on the run that the Japanese like: for instance, during the endless performances of Kabuki theatre the audience opens crackling cellophane containers and uses chopsticks to extract mouthfuls of white rice and raw fish.)

What is the relationship between the girl and the woman? A niece, a maid, a lady-in-waiting? She is always on the move, coming and going, chattering with total spontaneity: now she's coming back from the restaurant-car, carrying a fresh drink. And the woman? It seems as if she is owed everything, she constantly has her nose in the air. It is at times like this that one feels the difference between two cultures, when one does not know how to define what one sees, the gestures and behaviours, when one is not able to tell what is usual and what is individual in them, what is normal and what is unusual. Even if tomorrow I were to try to ask a Japanese person willing to listen: 'I saw two people like this and this. Who might they be? What is their social or family relationship?', I would find it difficult to make my curiosity understood, or get appropriate answers, and in any case every definition of a role would require the explanation of the context in which that role is found, it would lead to new questions and so on.

Outside the window an endless suburb rolls past. I glance at the headline in the *Japan Times*, an English-language daily. Today is the fiftieth anniversary of the coronation of the Japanese Emperor, and the government has declared a solemn celebration. There have been many polemics about the appropriateness of such a ceremony: the left is against it; protest demonstrations are planned; there are fears of a terrorist attack. For some days now the police have been guarding every crossroads in Tokyo; the nationalist associations' vans, blaring out military songs, have been criss-crossing the city, which is bedecked with flags.

That morning, during my taxi-ride from the hotel, Tokyo was full of lines of policemen, holding their riot-shields and long truncheons. In an empty lot about a hundred young people were sitting on the ground amidst their red flags, underneath a booming loudspeaker: this was clearly one of the protest meetings that had been organized across the city.

(Quick impressions of my first days in Tokyo. This is a city that is all elevated highways, flyovers, monorail trains, junctions, queues of traffic moving slowly on different levels, underpasses, underground pedestrian tunnels: a metropolis where everything can happen at the same time, as though in dimensions that do not communicate with, and are indifferent to, each other. Every event is circumscribed, constituting an order on its own, which is marked off and then enclosed in the surrounding orderliness. In the rainy evening atmosphere a strike demonstration goes by, channelled into one of the lanes on the road: it stops at a traffic light, then starts off again at green, keeping time with the sounds of the whistles, carrying totally identical red flags, preceded and followed by a squad of policemen, as though in parenthesis, while the traffic continues to go by in the other lanes. Everybody stares in front of themselves, never to the side.)

The *Japan Times* has interviewed about fifty Japanese celebrities (mostly artists and people from the world of sport) on their feelings about the Emperor and about the celebrations. As far as the festivities are concerned, many are indifferent or doubtful; on his person and the institution opinions vary from unconditional reverence (especially amongst the older interviewees) to still emotional memories of when they first heard the voice of this being who had up until then been invisible and unapproachable (when he announced Japan's surrender on the radio, a month after the atomic bombs), to perplexity at such a long reign on a throne that is purely symbolic. (The Emperor is something more and something less than a constitutional monarch: according to the constitution he is the 'symbol of the state and of the unity of the people' but he lacks any power or function.) 'About half of these fifty years of rule have involved wars and invasions,' is the memory of one elderly writer, who says he is against the celebrations while still maintaining his respect for the person and the institution.

(That evening, the television shows the images from the day in Tokyo, which are very clear even for those who do not understand the words of the commentator. In a series of rapid shots the 'serpent' of demonstrators snakes by, swaying with their heads held low; the police advance with riot-shields and truncheons raised;

there are police charges, a melee, a hail of kicks against someone crouching on the ground; then longer sequences of celebrations: children with flowers, little flags, lanterns. In a huge hall the tiny Emperor, dressed in tails, reads out his speech, his bespectacled eyes running through the lines from top to bottom; the Empress is seated beside him, in a light-coloured dress and hat. In his speech – according to the headlines in next day's paper – the Emperor declares himself sorry for the victims of the Second World War.)

In the first days in a new country one makes the effort to establish links between everything one sees. In the train my attention is divided between reading the comments on the Emperor and observing the old, impassive lady, being treated with due reverence and respect in the middle of a train full of businessmen who sit with their paperwork on their knees: budgets, estimates or blueprints for machines and buildings.

In Japan invisible distances are more powerful than visible ones. In Tokyo a central street runs alongside the canal which surrounds the green area of the imperial palaces. The endless traffic-jam laps against this boundary, beyond which everything is silence. The gates of the gardens open to the crowds just twice a year, but the whole year round groups of pilgrims get off coaches and head off on foot following a tour-guide with a little flag along the walls up to the gates of the Square of the Double Bridge, where they have their group photograph taken. This is the final frontier that common mortals are allowed to reach on normal days; beyond begins the rulers' residence, which is like an area beyond the bounds of the earth. I went there too, like the diligent tourist I am, but it was impossible to see anything at all: guards on duty, a twin-arched bridge over the canal, amidst the weeping willows.

The young girl has now sat down beside the woman, and is talking and laughing. The woman stays silent and stern-faced, does not reply, does not turn round, stares straight in front of her. The girl continues to chatter cheerfully, gently, as though hopping from one subject to another, improvising some lines of talk or trying out some jokes, putting into practice an art of conversation that is well defined and discreet, a rule of behaviour that is second nature to her and spontaneous, almost as if she were performing musical variations on a keyboard. And the old woman? Silent, serious, dour. She is not necessarily not listening: but it is as if she were sitting by the radio, receiving a communication that did not require any reply on her part.

In short, this old woman is ghastly and unpleasant! She is arrogant and selfish! A monster! Even those who, like me, try to abstain from formulating opinions on what they are not sure of understanding can be subject to sudden outbursts of anger. So at this point I am seething inside against the old woman who seems to me to be the embodiment of something terribly unfair. Just who does she think she is? How can she claim to deserve so much attention? My resentment towards the woman's arrogance increases along with my admiration for the girl's grace and happiness and civil attitude – qualities which are for me equally mysterious – and I feel that the way they are spurned is unforgivable.

If I consider the matter more carefully, it is a complex and mixed state of mind that is nagging away at me at this stage. There is certainly the urge to rebel aroused by my solidarity with the young against the crushing authority of the old, with the downtrodden against the privilege of lords and masters. There is all this, certainly. But perhaps there is also something else: a hint of envy, a rage that stems from my somehow identifying with the position of the old woman, the desire to tell her through clenched teeth: 'Don't you know, you fool, that where we come from, in the West, it will never again be possible for anyone to be waited on as you have been? Don't you know that in the West no old person will ever be treated with so much devotion by the young?'

Actually it is only by representing the conflict as something happening inside myself that I can have any hope of penetrating the secret, of deciphering it. But is that really the case? What do I know of life in this country? I have never been inside a Japanese house, and this is the first time (and it will be the last) during my trip that I have had the chance to glimpse something like a scene from domestic life.

The thin doors of the traditional Japanese house might seem to slide apart like curtains opening on a stage that holds no secrets. But actually this is not the case: this is a world where inside and outside are separated by a psychological barrier that is difficult to cross. The proof of this is in their pictorial representations. It was in the West, in the fourteenth century, that painters solved once and for all the problem of representing interiors in a way that seems obvious to us today, in other words by abolishing one wall and showing the room opened up like a stage scene. But a couple of centuries earlier Japanese painters of the twelfth century had found another system, less direct but more complete, of exploring visually the interior space while respecting its separation from the outside: they abolished the roof.

In the painted hand-scrolls that illustrate the manuscripts of the refined Court literature of the Heian period, the style known as *fukinuki-yatai* (which means precisely that, 'house without roof') shows us stylized characters, with no depth, in an oblique geometrical perspective of partitions, door-frames, walls that are only as high as screens. This allows us to see what is going on at the same time in various rooms.

Every time I happen to glance over the seat-back that separates me from the two women, the scene changes: now it is the old woman who is talking, in a patient, measured fashion. That's it: there now seems to be a perfect understanding between the two.

A few days previously, in a Tokyo museum, I had stopped to look at some of the very elegant rolls that illustrate the diary and novel of the exquisite Murasaki. Now the presence of the young woman with her full smile and the gentle and composed poses she makes with her neck, shoulders and arms, like a character from Murasaki in the midst of a world of harshness, makes this interior of the electric train seem like one of those roofless houses that reveal and at the same time conceal views from a secret life on a painted hand-scroll.

The Obverse of the Sublime

In November the maple trees' leaves turn a scarlet red, which is the dominant note of the autumn landscape in Japan, standing out against the dark green background of the conifers and against the various shades of tawny, rust and yellow of the rest of the foliage. But it is not with an act of outrageous chromatic arrogance that the maples impose themselves on one's view: if the eye is drawn towards them as though lured by a musical motif, it is because of the lightness of their starry leaves, which seem suspended around their thin branches, the leaves all horizontal, without depth, yearning to expand and at the same time not to clutter the transparency of the air.

The leaves of the ginkgo tree are yellow, a very sharp and luminous yellow, and they fall like rain from the highest branches like flower petals: infinite numbers of little leaves the shape of fans, a constant light rain that turns the surface of the little lake yellow.

The guide is explaining in Japanese the history of the Sento imperial palace to a group of visitors: it was built in the seventeenth century for the ex-Emperors in a period when both voluntary and forced abdications were frequent, as all power was in the hands of the generals. The imperial villas of Kyoto can be visited only with a special permit which has to be applied for in writing. For foreigners the waiting time for the permit can be just a few days, but for the Japanese themselves it takes at least six months, and to have visited these historically famous places is a piece of good fortune that does not come everyone's way. Every person hoping to make a visit is summoned on a certain date, and is assigned to a group which a guide will lead along the prescribed itinerary, stopping at pre-arranged points for explanations in Japanese or English, depending on the make-up of the group. I know too little about the dynastic history of Japan to benefit from the guide's talks; I expect to profit more from the moments of pause, from small deviations from the group's itinerary, from people and details I come across by chance.

An old woman goes by, very tiny, dressed in purple, with her head shaved, certainly a nun; she is wizened and bent almost double. Many old women in Japan are hunchbacked and twisted as if related to the dwarf trees cultivated in pots according to the ancient art of *bonsai*.

Even the shape of the tall trees is the result of careful pruning. Here are two gardeners pruning pine-trees, climbing up on triangular steps which have a bamboo supporting pole. It seems as if they are fleecing every branch-top with their fingers, leaving only a horizontal tuft, so that the tree's crown spreads out like an umbrella.

The majority of the gardeners are women. A team of them advances along the path, dressed in what must be the traditional working outfit: blue trousers, a grey blouse, a kerchief on their head. They are low in stature beneath great bundles of dry leaves and baskets of branches and are armed with rakes and pruning hooks; it is impossible to say if they are young or old, but they are already knotted and contorted, as though adapting to the environment.

There is one thing I seem to be starting to understand here in Kyoto: something I've learned through the gardens more than through the temples and palaces. The construction of a nature that can be mastered by the mind so that the mind can in turn receive a sense of rhythm and proportion from nature: that is how one could define the intention that has led to the layout of these gardens. Everything here has to seem spontaneous and for that reason everything is calculated: the relationships between the colours of the leaves in the various seasons, between the masses of vegetation depending on their different times of growth, the harmonious irregularities, the paths that climb up and descend, the pools, the bridges.

The little lakes are an element in the garden that is just as important as the vegetation. There are usually two of them, one of flowing water, the other a still pool, which create two different landscapes, to tone in with two different states of mind. The Sento garden has two waterfalls as well: a male and female waterfall (Odaki and Medaki), the first high over the rocks, the second one murmuring as it rushes between steps made of little stones in a gap in the lawn.

The lawns have moss rather than grass. There is a moss that is made up of actual little plants a few centimetres high; in Japanese it is called cedar-moss because these little plants resemble minuscule conifers. (There is a temple in Kyoto whose garden is entirely covered in moss: you can count a hundred different kinds of moss there; or at least thirty, if you use more rigorous classifications. But with this temple of moss one enters a different world: it is as if one were entering a Nordic park soaked with rain. Actually every characterization that is too extreme takes us away from the true spirit of Japanese gardens, where no element ever takes precedence over another.)

Every aspect of the garden is designed to arouse admiration, but using only the simplest means: these are all familiar plants – no seeking after sensational effects. Flowers are almost absent; there are a few white and red camellias; it is autumn, and it is the leaves that supply the colour; but flowering plants are also absent; in spring it will be the fruit trees that will blossom.

Hillocks, rocks, slopes multiply the landscape. Groups of plants are arranged according to their mutual proportions in order to create illusions of perspective: backdrops of trees which seem to be distant are actually nearby; views of rises or descents suggest spaces that are not actually there. The Japanese passion for the small that provides the illusion of the big comes out also in the composition of the landscape.

I am being accompanied on my visit to Kyoto by a Japanese student who is a passionate reader of poetry and a poet himself: he reads Italian very well and speaks it a little too. But conversation is difficult because both of us would like to say things that are either too precise or too nuanced, and instead all we manage is to come up with statements that are either too generic or too peremptory.

The young man explains that, before being frequented by Emperors, these places were popular with famous poets, who are now commemorated by plaques and little temples amidst the trees. Following the line of my reflections, it occurs to me that poetry and gardens generate each other in turn: the gardens were created as illustrations for poems and the poems were composed as a commentary on the gardens. But I come to think of this more from my love of symmetry in my thoughts than because I am really convinced of it: or rather, I find it very plausible that one can make the equivalent of a poem with the way one arranges trees, but I suspect that real trees are of little or no use for writing a poem about trees.

Suddenly I see, standing out above the red, rust-coloured and yellow trees beyond the lake, the bare branches of a single tree that has lost its leaves. Amidst that blaze of colours those black, dead branches make a funereal contrast. A flock of birds flies past, and amidst all the other trees around they home straight in on the bare tree, swoop down on to its branches, landing there one by one, black against the sky, enjoying the November sunshine.

I think: right then, the landscape has given me the subject for a poem; if I knew Japanese, all I would need to do is describe this scene in three lines consisting of seventeen syllables in total, and I would have composed a haiku. I try conveying this idea to the young poet. He does not seem convinced. A sure sign that haikus are composed in a different way. Or that it makes no sense to expect a landscape to dictate poems to you, because a poem is made of ideas and words and syllables, whereas a landscape is composed of leaves and colours and light.

The rooms of the imperial palace, constantly destroyed and rebuilt over the course of the ten centuries that the Court resided at Kyoto, can be seen from the outside, through the open sliding doors, like a theatre stage. One mat that is higher than the others on the floor marks the place reserved for the Emperor. The Japanese house, the royal palace included, is a series of empty rooms and corridors, with mats instead of furniture, no chairs, beds or tables: a place where no one ever stands or sits, where people only crouch or are on their knees, with few objects placed on the ground or on low stools or in niches, such as a vase with a few branches in it, or a painted screen.

All trace of life seems to have been removed from this model house; there is none of the heavy weight of existences that materializes in our furnishings and impregnates all our Western rooms. Visiting the Court palaces in Kyoto or those of the great feudal landowners, one finds oneself wondering whether this aesthetic and moral ideal of the bare and unadorned was achievable only at the peak of authority and wealth, and whether it presupposed other houses chock-full of people and tools and junk and rubbish, with the smell of frying, sweat, sleep, houses full of bad moods, people rushing, places where people shelled peas, sliced fish, darned socks, washed sheets, emptied bed-pans.

These Kyoto villas, whether they were inhabited by sovereigns still in power or by those who were retired, convey the idea that it is possible to live in a world separate from what constitutes the real world, sheltered from the catastrophes and incongruities of history, a place which reflects the landscape of the wise man's mind, free from all passions and neuroses.

Crossing over the Six-Slab Bridge, made of curved slabs of stone, and going along a path amidst the multi-coloured foliage of the dwarf-bamboo plants, I try to imagine myself as one of the former Emperors of an empire that was at the mercy of the whims and devastation carried out by the lawless landowners, perhaps cheerfully resigned to concentrating on the one operation that is still possible for him: contemplating and guarding the image of how the world should be.

Deep in these thoughts, I had wandered away from the group of visitors, when out from behind a hedge popped a custodian with a walkie-talkie who sent me back to the ranks. Wandering around the garden on your own is not allowed. Blending with the swarm of tourists who open their camera lenses at every panoramic view, I can no longer create the distance necessary for contemplation. The garden becomes an indecipherable calligram.

'Do you like all this?' asked my student. 'I cannot help thinking that this perfection and harmony cost so much misery to millions of people over the centuries.'

'But isn't the cost of culture always this?' I object. Creating a space and time for reflection and imagination and study presupposes an accumulation of wealth, and behind every accumulation of wealth there are obscure lives subject to labour and sacrifices and oppression without any hope. Every project or image that allows us to reach out towards another way of being outside the injustice that surrounds us carries the mark of the injustice without which it could not have been conceived.

'It is up to us to see this garden as "the space from another history", born from our desire for history to obey other rules,' I say, remembering I had recently read an introduction to Petrarch's *Canzoniere* by Andrea Zanzotto in which this idea is applied to Petrarch's poems. 'We should see it as a project for finding a different space and time, a proof that the total domination of sound and fury can be challenged . . .'

The group has reached a bed of smoothed, round stones, bright grey and dark grey in colour, which continue beneath the green water of the little lake as though revelling in its transparency.

'These stones,' the guide was explaining, 'were brought here three centuries ago from every part of Japan. The Emperor rewarded whoever brought him a bag of stones with a bag of rice.'

The student shakes his head and looks bitter. We seem to see the queue of peasants conjured up by those words bent double under their bags of stones, snaking across the little bridges and paths. They deposit the loads they have carried from distant regions in front of the Emperor, who examines the stones one by one, places one in the water, another one on the side of the lake, and rejects many others. Meanwhile the attendants busy themselves round the scales: on one dish there are the stones, on the other rice . . .

The Wooden Temple

In Japan that which is a product of art does not hide or modify the natural elements from which it is made. This is a constant feature of the Japanese spirit which their gardens help us to understand. In buildings and traditional objects, as also in their cuisine, the materials from which these things are made are always recognizable. Japanese cooking is a composition of natural ingredients but one that is aimed primarily at a visual effect, and these elements reach the table retaining their original appearance to a large extent, without having undergone the metamorphoses of Western cuisine, where a dish becomes more a work of art the more unrecognizable its ingredients are.

In their gardens the various elements are put together according to criteria of harmony and criteria of meaning. The only difference is that these vegetal equivalents of words change shape and colour over the course of the year and even more so over the course of years: these complete or partial changes were factored in when the garden-poem was planned. Then the plants die and are replaced by others that are similar and are laid out in the same places: as the centuries go by the garden is continually remade but always remains the same.

And this is another constant highlighted by these gardens: in Japan antiquity does not have its ideal material in stone as in the West, where an object or building is considered ancient only if it is conserved in its substance. Here we are in the universe of wood: what is ancient here is that which perpetuates its design through the continual destruction and renovation of its perishable elements. This holds for gardens as it does for temples, palaces, villas and pavilions, all of which are in wood, all destroyed many times by the flames of fire, many times covered in mould and rotten or reduced to dust by woodworm, but refashioned piece by piece every time. The roofs made of layers of pressed cypress bark are remade every sixty years, as are the trunks forming pilasters and beams, the walls made of planks, the bamboo ceilings, the floors covered in mats (the omnipresent *tatami* mats, which act as a unit of measurement of interior surfaces).

During the visit to Kyoto's centuries-old buildings the guide points out how often they take care to replace this or that piece of the construction: the fragility of its parts emphasizes all the more the antiquity of the whole. Dynasties, human lives, the fibres of tree-trunks rise and fall, but what lasts is the ideal shape of the building, and it does not matter if every piece of its structural support has been removed and replaced countless times, and the most recent replacements smell of newly planed wood. In the same way the garden remains the garden designed 500 years ago by a poet-architect, even though every plant follows the course of the seasons, rains, frosts, wind; similarly the lines of a poem are handed down over time while the paper of the pages on which the lines are systematically written disappears into dust.

The wooden temple marks the junction of two dimensions of time: but in order to understand it we need to remove from our minds phrases such as 'being and becoming', because if everything is reduced to the language of philosophy from the world which we have started out from, it was not worthwhile coming all this way on my journey. What the wooden temple can teach us is this: in order to enter into the dimension of continuous, single and infinite time the only way is to go through its opposite, the perpetuity of the vegetal, the fragmented and ramified time of that which is replaced, is disseminated, buds, dries up or wastes away.

Rather than the temples full of statues with their high, pagoda-like structures, I am attracted more by the low constructions and interiors furnished only with mats, which are usually secular buildings, villas or pavilions, but also sometimes temples or sanctuaries that invite one to abstract meditation, to disembodied concentration.

Such is the temple called the Silver Pavilion, a supple wooden construction on two floors beside a small lake, with just one statue (of Kannon, the female incarnation of the Buddha) in a space for Zen meditation called the Hall of Emptied Mind. Such also is the temple called Manju-in, which an incompetent like myself would swear is Zen but actually is not: this is a temple which seems like a villa, with many low rooms, almost empty, with their *tatami*, the *ikebana* vases (which at this time of year contain pine boughs and camellias, bird-of-paradise flowers and camellias, and other autumn combinations), and a few unobtrusive statues, and many small gardens all around.

The wooden temple reaches its perfection the more the space that welcomes you is bare and unadorned, because all that matters is the material in which it is built and the ease with which one can undo it and reassemble it exactly as it had been before, in order to prove that all the bits of the universe can fall one after the other but there is something that remains.

The Thousand Gardens

A path made of irregular stone slabs snakes its way around the full length of the imperial villa of Katsura. As opposed to the other gardens in Kyoto made for static contemplation, here inner harmony is reached by following the path step by step and reviewing each image that your sight perceives. If elsewhere a path is only a means to an end and it is the places it leads to that speak to the mind, here the footpath is the *raison d'être* of the garden, the main theme of its discourse, the sentence that gives meaning to its every word.

But what meanings? The path on this side of the gate is made of smooth stones but on the other side it is made of rough ones: is this meant to be the contrast between civilization and nature? On the other side the path splits into a straight branch and a twisted one; the former comes to a dead end, while the latter continues: is this a lesson about how one should move in the world? Every interpretation leaves one dissatisfied: if there is a message, it is the one we grasp in sensations and things, without translating them into words.

These stones are embedded in moss, and are flat, separate from each other and set at the right distance so that the person walking always finds one under his foot at each step taken; and it is precisely because they correspond to the measure of our footsteps that the stones actually dictate the movements of the person walking, force us into a calm, uniform pace, determining both the route and its halts.

Each stone corresponds to a footstep, and at each step there is a landscape that has been studied down to the last detail, as in a painting. The garden has been so set out that at each step one's gaze meets different perspectives, a different harmony in the distances that separate the bush, the lantern, the maple tree, the hump-backed bridge, the stream. As one moves along the route, the scenery changes totally many times, from thick foliage to a clearing dotted with rocks, from the lake with the waterfall to the lake of standing water; and each scenario in turn is broken down into views that take shape as soon as one moves: the garden multiplies into endless gardens.

The human mind has a mysterious mechanism whereby we are convinced that that particular stone is always the same stone, even though its image – at the slightest movement of our gaze – changes shape, dimensions, colour, outlines. Every single or limited fragment of the universe breaks up into an infinite multiplicity: all you have to do is go round this low stone lantern and it turns into an infinity of stone lanterns; this fret-worked polyhedron of stone, marked with lichens, becomes doubled and quadrupled and sextupled, turning into a totally different object depending on which side you look at it from, on whether you are approaching or leaving it.

The metamorphoses generated by space are in addition to those caused by time: the garden – each of the infinite number of gardens – changes with the passing of the hours, the seasons, the clouds in the sky. The Emperors who designed Katsura arranged for platforms of bamboo canes to be made in order that visitors might see the blossoming of the peach trees in April, or the reddening of the maple leaves in November, and set up four teahouses, one for each season, each of which looks out on an ideal landscape at a particular point in the year; each ideal landscape in a season has in turn a time in the day or night that is its ideal moment. But there are four seasons and the hours run their cycle around midday and midnight. With its recurring moments time removes the idea of the infinite: this is a calendar of exemplary moments which are repeated cyclically, and which the garden tries to fix in a certain number of places.

What about space, then? If there is a correspondence between the points of view and our footsteps, if every time our right or left foot advances on to the next slab a perspective opens up that has been decided by the person who designed the garden, then the infinity of viewpoints is reduced to a finite number of views, each one separate from the one that precedes and follows it, and characterized by elements that distinguish it from the others: a series of precise models each one of which corresponds to one necessity and one intention. So here is what the path is: of course it is a device for multiplying the garden, but also for removing it from the vertigo of the infinite. The smooth slabs that make up the path at the villa of Katsura are 1,716 in number – this figure, which I found in a book, seems likely to me considering that there are two slabs per half metre for a total distance of half a mile – so one can go through the garden in 1,716 steps and contemplate it from 1,716 perspectives. There is no reason to be seized by anguish: that clump of bamboo can be seen from a precise number of different perspectives, no more and no less, the chiaroscuro varying according to how thinly or densely the stems are clustered together, and one feels distinctly different sensations and feelings at each step, a multiplicity that now I feel I can master without being overcome by it.

Walking presupposes that at every step the world changes in some aspect and also that something changes in us. Consequently, the ancient masters of the tea-ceremony decided that in order to reach the pavilion where tea will be served the guests should walk along a path, pause at a bench, look at the trees, pass through a gate, wash their hands in a basin dug out of a rock, follow the path marked out by smooth slabs all the way to the simple hut which is the tea-pavilion, to its very low door, where everyone has to bow down to enter. In the room, there are only mats on the floor, a stool with very sophisticated cups and tea-pot, a recess in the wall – the *tokonoma* – where an exquisite object is displayed, or a vase with two sprigs in flower, or a painting, or a sheet of paper with rows of calligrams. It is by limiting the number of things around us that one prepares oneself for accepting the idea of a world that is infinitely larger than ours. The universe is an equilibrium of solids and voids. The words and gestures that accompany the pouring of the foaming tea must have space and silence around them, but also a sense of inner meditation, of a limit.

The art of the greatest master of the tea-ceremony, Sen-no Rikyu

(1521–91), which was always based on the maximum simplicity, expressed itself also in the design of the gardens around the tea-houses and temples. Interior events present themselves to one's consciousness through physical movements, gestures, journeys, unexpected sensations.

A temple near Osaka had a wonderful view over the sea. Rikyu had two hedges planted which totally hid the landscape, and near them he had a small stone pond built. Only when a visitor bent over the pond to take water in the hollow of his hands would his gaze meet the oblique gap between the two hedges, and then the vista of the boundless sea would open up before him.

Rikyu's idea was probably this: bending down over the pond and seeing his own image shrunk in that narrow stretch of water, the man would consider his own smallness; then, as soon as he raised his face to drink from his hand, he would be dazzled by the immensity of the sea and would become aware that he was part of an infinite universe. But these are things that are ruined if you try to explain them too much. To the person who asked him about why he had built the hedge, Rikyu would simply quote the lines of the poet Sogi: 'Here, just some water, / There amidst the trees / The sea!' ('*Umi sukoshi / Niwa ni izumi no / Ko no ma ka na!*')

The Moon Chasing the Moon

In the Zen gardens of Kyoto there is a white, coarse-grained gravel which has the power to reflect the moon's rays. At the Ryoanji temple, this sand, raked by the monks into straight parallel furrows or into concentric circles, forms a little garden around five irregular groups of low rocks. At the temple of the Silver Pavilion, on the other hand, the sand is arranged into a circular mound, on its own, like an upturned cone, and stretches out in an expanse that is raked in regular waves. Beyond it a lively garden of bushes and trees extends around a little lake that has a wild look to it. On the nights when there is a full moon, the whole garden is illuminated by the silver sparkle of the sand. I visited the Silver Pavilion only in the daytime, and it was raining; but those rain-soaked white grains seemed to return the lunar light which they had stored; a mirror image of the source of that light seemed to be stored in those shapes in the white sand, in that volcano which seemed as sodden as a sponge, under the raindrops falling straight as moon-rays on to the raked parallel tracks that a monk reshapes every morning.

Love for the moon often has its double in love for its reflection, as if to stress a vocation for mirror games in that reflected light. Of the four tea-houses of the sixteenth-century Katsura Villa in Kyoto – one for every season, each arranged differently and characterized by different landscapes – the autumnal one is sited in such a way as to allow you to see the moon at the moment it rises and to enjoy its reflection in the lake.

This fascination for duplication, typical of the image of the moon, is probably the source of a poem by a curious poet from the early twentieth-century Japanese avant-garde, Tarufo Inagachi.

Even in a word-for-word translation, this poem seems to let us intuit (as in a reflection, appropriately enough) something of the fantasy that triggered it. It is called 'The Moon in Its Pocket'.

One evening the moon was walking down the street, carrying itself in a pocket. As it went down the hill, one of its shoelaces came undone. The moon bent down to tie the shoelace and the moon fell out of its pocket and started to roll quickly down the tarmac road that was soaking wet from the sudden shower. The moon chased after the moon, but the distance between them increased, thanks to the acceleration of lunar gravity as it rolled along. And the moon lost itself in the blue haze down there at the bottom of the slope.

The Sword and the Leaves

At the National Museum in Tokyo there is an exhibition of arms and armour from ancient Japan. The first impression it makes on you is that the helmets, breastplates, shields and broadswords had as their main purpose not that of defending or striking anyone but that of terrifying adversaries, imposing an image on them that would strike terror in their hearts.

The war-masks, contorted into cruel and threatening grimaces, sit between helmets adorned with horns, fins and griffin-wings, and sumptuous breastplates that inflate the chest with all their loops and spikes.

Those who are like me in that, when they visit Renaissance armouries in the West, they feel the pleasant, classic detachment of a reader of chivalric poems (the great cavalcade that is the armour room in the Metropolitan Museum in New York is for me one of the wonders of the world), here for the first time do not think of these artefacts as fantastic toys but consider more the message the objects were meant to transmit *in situ*; in other words they look at them just as today we would look at an armoured car on a battlefield. My reaction is immediate: I start running.

I run through room after room full of cases where countless sword-blades are exhibited, or different kinds of curved sabres, made of shining, tempered iron, razor sharp, with no handles, each one resting on a white cloth. Blades and blades and blades that all seem the same to me, and yet each of them has a label with long explanations. Crowds of people stop in front of every case, observing sword after sword with an attentive, admiring gaze.

Most of the visitors are men; but it is Sunday, the museum is crowded with families, and there are also ordinary women and children contemplating these swords. What do they see in those grim unsheathed blades? What do they find fascinating in them? My visit to the exhibition is carried out almost at running pace; the shine of steel transmits a sensation that is more auditory than visual, like swift hisses slicing through the air. The white cloths inspire a kind of surgical horror in me.

And yet I am well aware that the art of fencing in Japan is an ancient spiritual discipline. I've read the books on Zen Buddhism by Dr Suzuki. I remember that the perfect Samurai must never concentrate his attention on his enemy's blade, nor on his own, nor on striking his opponent, nor on defending himself, but must only annihilate his own ego; that it is not with the sword but with the non-sword that victory is won; that the master sword-makers reach the peak of their art through religious *askesis*. I know all this very well: but it is one thing to read something in books, quite another to understand it in real life.

A few days later I find myself in Kyoto: I walk through the gardens that had once been the haunt of exquisite poets, emperor-philosophers, hermit-monks. Amidst the hump-back bridges over the streams, the weeping willows that are reflected in the ponds, the moss lawns, the maples with their red star-shaped leaves, suddenly what comes back to my mind are the warrior masks with their terrifying grimaces, the looming approach of those giant warriors, the sharp edge of those blades.

Looking at the yellow leaves falling into the water, I remember a Zen story which only now do I think I understand.

The pupil of a great sword-maker claimed to have outdone his master. To prove how sharp his sword-blades were he immersed a sword in a stream. The dead leaves carried down by the current were neatly sliced in two as they went across the blade's edge. The master plunged into the stream a sword that he had fashioned. The leaves flowed on, slipping right past the blade.

The Pinballs of Solitude

In Tokyo and every Japanese city the word *Pachinko* written in the Latin alphabet means the pinball or electric billiard arcades, which are different from those in America or Europe in that the machines are vertical, arranged on the wall in a row, one next to the other, and you play them sitting down.

Judging by the number of arcades and the crowds in them at all times of the day and night, one would say that *pachinko* is the great Japanese passion of today. The arcades are decorated in the colours of the rainbow, inside and out, and illuminated by neon tubes and coloured lights that flash on and off. The tinny music coming out of the speakers matches this visual glitz. But were it not for this chromatic and acoustic aggression we would not notice that this is a place of entertainment, seeing these rows of people sitting on stools, everyone opposite their vertical window as though it were their place of work, their eyes staring at the flicks of the shiny machine, working the handles like a robot. The impression one comes away with is that of a factory floor, or of an office full of electronic devices, at its busiest hour.

In the West pinball machines in bars and in arcades are nearly always surrounded by groups of youngsters busy challenging each other and betting and taking the mickey. Here the impression one has is of a crowded solitude: nobody seems to know anyone else, everyone concentrates on their own game, staring into that flashing labyrinth and ignoring their neighbour to right and left; everyone is walled in, as it were, inside their invisible cell, isolated in this obsession or curse.

You can find *pachinkos* nearly everywhere, in the various centres of the polycentric city that is Tokyo as well as in its different suburbs, but above all in the nightlife districts. In the midst of night-clubs, pizzerias with their Italian colours, strip-clubs, bars, *poruno-shops* (the word 'porno' is adapted to conform to Japanese pronunciation), surrounded by the smell of eel that is either raw or fried in soya-oil, in the midst of this noisy world the *pachinkos* open up like metallic gardens offering a haven for the individual wanting to do something that will fully absorb his attention.

The players are mostly men, of all ages; but in the morning, when the signs in the entertainment areas have all gone out, only the *pachinkos'* rainbows stay lit and a new public takes over the little billiard balls: respectable housewives with their shopping baskets. Middle-aged or elderly women, mostly, with their garish-coloured kimonos, and large bows tied at their backs, their clogs over their white socks, sit down at these machines, placing by their side shopping-bags with celery and sweet potatoes sticking out, and very quickly, as though working a sewing-machine or electric loom, they devote their calm and contented attention to the bounces of the little balls.

Nightlife in Tokyo extends across several districts: Ginza, Shibuya, Shinjuku, from the most elegant to the most popular areas. You could almost say that half of the metropolis has no other purpose than to entertain the other half.

The crab restaurants are topped by signs that could be seen as extraordinary works of Pop Art: a giant crab occupying the whole of the façade of the building moves its legs and claws in all their different articulations, its protruding eyes going up and down in a constant rhythm. But the most sumptuous façades are those of the cafés, which are considered the *ne plus ultra* of Westernness. And what is more Western than an English castle? So then the cafés – usually on two or more floors – have a façade that portrays a medieval manor and have names which, in order to create an English atmosphere, rejoice in tautological titles such as 'The Mansion House'.

The miracle that everyone talks about and that never ceases to amaze people is that this over-populated metropolis has very low crime levels: violence is very rare, and women can go out alone at any hour of the day or night even in these districts without being molested (except by the odd drunk).

It is true that nightlife finishes early: at midnight all the places close because that is what is laid down by the law of this country, which has always practised austerity. (Only the night-spots classified as 'private clubs', in other words very expensive clubs, remain open.) The other reason is the transport problem. Everywhere starts to empty already by 10 p.m., night-clubs and pizzerias, cinemas and *pachinkos*, because the vast majority of the people live in distant suburbs and they have two hours of travel still to go; they cannot miss the last metro or train, and they have to go to sleep early in order to be up at dawn the next day to face two more hours on the train in order to go to work.

Eros and Discontinuity

Some observations on Japanese erotic prints. In them the human body appears to be shaped by three distinct elements:

1. faces intensely concentrated, engrossed in a kind of inward gaze;
2. bodies whose outlines are drawn with calm, clean lines and with colourless surfaces conjuring up pale skin and soft, muscle-free flesh – with no difference between men and women;
3. sexual organs represented with a technique that is much more detailed, a three-dimensional rendering (drawn with many lines and dark colouring) which manages to show everything: pubic hair, the labia maiora, sometimes even the inside of the woman's vulva, and the male member like a turgid entrail: this is a stylistic departure from the rest of the drawing and reveals in the sexual organs both a nature that is completely different, independent from the rest of the person and a savage ferocity.

The discontinuity of these aspects is stressed by the fact that the bodies are partially draped with clothes or blankets, hiding the details of the tangle of limbs entwined and superimposed on each other, so much so that the first action in our 'reading' of them, which is anything but instantaneous, is that of recognizing to whom this or that limb belongs.

Such stylistic eclecticism seems to have been created deliberately to convey the simultaneous presence and action of very different aesthetic and emotional factors in physical love.

The Ninety-Ninth Tree

The history of every temple and palace is interwoven with dynastic events and the preachings of Buddhist sects. As I go around I hear snatches of guides and escorts intoning flat, readily forgettable facts. The student who is acting as my interpreter condenses whole stories into one sentence that is comprehensible but lacking in emotional appeal. And yet those stories were recounted in a fascinating, warm, exciting way by the taxi-driver, who unfortunately only speaks Japanese.

The taxi that has been put at this guest's disposal during his stay in Kyoto is driven by a small, round, dynamic, giggly man, Mr Fuji, who takes his white-gloved hand off the gear-stick (Japanese taxi-drivers always wear white gloves) to indicate the main points about the places we have come through and which evoke famous episodes, and to bring them alive with enthusiastic gestures. He is the one who knows the whole history of this ancient capital, of the Courts which resided here and in nearby Nara for twelve centuries; he is the encyclopedia of local lore, but also the bard, the rhapsode of a world that has disappeared, buried beneath the thick covering of the present.

The taxi drives through an endless suburb of parking-lots, super-markets, warehouses, petrol-pumps (on which well-known brand-names pop up between indecipherable letters), factory sheds, baseball fields, rows of shops, used-car lots, electric pinball arcades. Only the maples that display their red leaves where you would least expect it, only the odd roof with the traditional concave wings, remind us that Japan is a 'different' country.

All of a sudden Mr Fuji gives a start, points to an invisible spot amidst the television aerials and says that on that spot one thousand years ago a palace stood, or a poet walked by the shore of a lake. The abyss that opens up between these scenes that he evokes and what one sees today does not seem to trouble him: the name connects the space with time, that point on a map that has now been abolished remains as the repository of a myth.

The story he tells me now is about an Emperor who fell in love with a very beautiful but haughty woman who lived over there (see behind that service station?). In order to test him, the lady said that he had to come to her one hundred times and declare his love to her, and only on the hundredth occasion would she consent to be his. The Emperor went to her every day, setting out from his distant palace (see, beyond that gasometer?), and every day he planted a tree in front of the haughty beauty's house. In this way he ended up planting ninety-nine trees. Just one more visit and the beautiful woman would be his . . .

At that point, having proved the constancy of his feelings, the Emperor decided to withdraw, to give up, and never appeared again. The trees grew into a wood, the Wood of the Ninety-Nine Trees, as it is called to this day.

One's gaze ranges over a horizon of cement and tarmac. But the taxi has turned down a little road amidst courtyards full of crates. Suddenly there's a tree, an enormous green tree of great height, of an unfamiliar species, but with myriads of tiny leaves. An old signpost states that this is the last surviving specimen from the Wood of the Ninety-Nine Trees, perhaps actually the ninety-ninth, thus proving that yesterday's geography of the sublime, so dear to Mr Fuji, really does have a link with today's geography of the prosaic, and that even today the roots planted in a terrain of risky investments with no capital guarantee still nourish the branches that face a world of balance sheets that must all be showing a profit, a world of operations that can never close at a loss.

Mexico

The Shape of the Tree

In Mexico, near Oaxaca, there is a tree that is said to be 2,000 years old. It is known as 'the Tule tree'. I have just got off a coach crowded with tourists, and, as I approach, even before my eye can make anything out, I am seized by a sense of threat: as though that vegetal cloud or mountain that is now outlined in my field of vision is sending out a warning that here nature, with slow, silent steps, is intent on furthering a plan of her own that has nothing to do with human proportions and dimensions.

I was just about to utter a cry of amazement, to compare what my eyes were seeing with the concept of tree which hitherto I had used to group together all the actual trees I had encountered, when I realized that what I was looking at was not the famous tree but another one of the same species growing nearby, clearly a bit younger and a bit less gigantic, seeing that the guide did not mention it. I turned round: suddenly I saw the real Tule tree there as though it had popped up just at that point. The impression it conveys is totally different from what I had been expecting. The almost spherical extent of its crown sitting above the gargantuan girth of the trunk makes the tree seem almost squat.

The Tule tree measures 40 metres in height, according to the guide, and 42 in perimeter. Its botanical name is *Taxodium distichum*; its Mexican name is the *sabino*.

It belongs to the cypress family but does not resemble a cypress at all: it is a bit like a sequoia, if that helps to give some idea. The tree towers above a church from the colonial period, Santa Maria del Tule, which is white with red and blue geometric ornamentation, like something out of a child's drawing. The church's foundations risk being undermined by the tree's roots.

Visiting Mexico, one finds oneself puzzling every day over pre-Hispanic ruins and statues and bas-reliefs, witnesses of an unimaginable 'before', of a world totally 'other' than our own. Then suddenly here we have a witness that is still alive today and was already living before the Conquista, in fact even before these plateaux saw the successive waves of Olmecs, Zapotecs, Mixtecs and Aztecs.

At the Jardin des Plantes in Paris I have always looked with wonder on the cross-section of a sequoia trunk from more or less the same period, which is displayed as though it were a compendium of world history: the great historical events of the last two thousand years are marked on little copper tags that are nailed to the concentric circle in the wood that corresponds to the year in question. But while the Paris tree is the relic of a dead plant, this one, the Tule tree, is a living thing, which barely shows any signs of effort in transporting lymph to its leaves. (In order to compensate for the aridity of the soil, they water the tree with injections of water to its roots.) This is certainly the oldest living thing I have ever come across.

I avoid the Japanese tourists walking backwards or crouching down, trying to cram this colossus into their lenses. I approach the trunk, and go round it to discover the secret of a living form that can resist time. My first impression is of an absence of form: this is a monster that grows – one might say – according to no plan; the trunk is both one and multiple, it is as if it were girdled by columns of other, smaller trunks that stick out all around the massive central trunk, or that detach themselves from it as if they wanted to make us believe that these are aerial roots descending from the branches like anchors trying to find the earth, whereas in fact they are proliferations of its earthly roots which have grown upwards. The trunk seems to embrace within its current perimeter a long history of uncertainties, repetitions, moments of branching out in different directions. Like boats which can't get out to sea, what stick out from the trunk are horizontal beams that were truncated a thousand years ago just as they were giving life to the bifurcation of the tree, and have lost all memory of that original intention, and have now just become short, stumpy protuberances. From the elbows and knees of branches which did survive catastrophe in remote times, rigid secondary branches continue to develop in what seems to be highly uncomfortable gesticulation. Knots and wounds in the wood have continued to spread, the former proliferating in bumps and accretions, the latter stretching their lacerated sides, emphasizing their peculiarity, like a sun around which a generation of cells radiates. And above all this, there is the continuity of the bark, which has thickened, developed calluses, grown on top of itself, and which reveals all the weariness of its decrepit skin and at the same time the immortality of something that has reached a condition that has so little life about it that it can no longer die.

Does that mean that the secret of survival is redundancy? Certainly it is by repeating its own messages countless times that the tree guarantees itself against the constant threat of fatal damage to its individual parts, and thus manages to impose and perpetuate its essential structure, the interdependence of roots, trunk and crown. But here we have gone beyond redundancy. What disturbs me as I go round the Tule tree is the willingness of morphology to change its roles, the disruption of vegetal syntax: roots rising upwards, segments of branches that have become trunk, segments of trunk that have been born from the bud on a branch. And yet the result, seen from a distance, is still always a tree – a super-tree – with its roots, trunk and crown all in the right place – super-roots, super-trunk and super-crown – as though the distorted syntax re-established itself at a higher level.

Is it through a chaotic waste of matter and forms that the tree manages to give itself a shape and maintain it? That means that the transmission of meaning is guaranteed in excessive display, in the profusion of self-expression, in throwing out matter by whatever means. Because of my temperament and upbringing I have always been convinced that the only thing that matters and survives is whatever is focused on one single end. Now the Tule tree proves me wrong, wants to convince me of the opposite.

My interview with the tree should begin now, but already the Japanese tourists have taken their pointless photographs and stopped swarming round the giant. I too have to take my seat in the coach that is setting off for the Mixtec ruins of Mitla.

Time and Branches

Also at Oaxaca, I see another extraordinary Mexican tree, but this time of painted stucco, in a seventeenth-century Dominican church. This is a decoration in relief of the vault of the church of San Domingo, reworking the motif of Christ's genealogical tree, the tree of Jesse (Jesse, father of David, from whose stock, according to the prophets, the Messiah was to be born). In art history this motif is often identified with that of the Tree of Life (in the latter case it starts with Adam and connects the Fall with the Redemption through the continuity of the wood of the Tree of Life and that of the Cross).

A thin, twisted trunk emerges from the body of a character lying supine and develops branches that cover the vault in a series of circles and a harmonic tangle of vegetal volutes, from which characters in relief stand out like grape-clusters on a vine branch (the plant also has clusters of real grapes, and vine leaves, which allow us to identify it as a vine). The coloured characters stand out against the white plaster: kings with golden crowns, bishops with mitres, warriors with armour and plumed helmets like Sicilian puppets, gentlemen with broad seventeenth-century collars. Apparently there are only a couple of female figures, one of whom is a nun. The top of the tree, towards which all the branches converge, supports a Madonna and child, surrounded by heads of angels.

It is not easy identifying the characters: if this really is meant to be a 'Tree of Jesse', then perhaps the forefather lying on his back is David, and one of the kings must be Solomon. But the figures are stereotypes and are dressed in a historically unspecific style somewhere between medieval and baroque, and also the order is probably arbitrary: according to the Gospels Christ's genealogy goes from father to son in a single line, whereas here the twisted trunk directly links the figure at the base with the figure at the top and all the other characters pop up at various heights on the lateral branches like generations of brothers. Unless, of course, the climbing motion of the plant requires us to read the line of succession in a freer way, following a serpentine movement.

According to some of the guides, however, the figure at the base is Saint Dominic and those on the branches are leading figures of the Dominican order (but in that case should they not all be wearing ecclesiastical garb?) whose faith converges on divine grace. Whatever the exact iconological interpretation might be, the sense of the tree's design is clear and of immediate visual efficacy: it has to connect a departure point with a point of arrival, both of them sacred and necessary, through an exuberance of forms of life which however also respond to a harmonic design, according to the intention of Divine Providence or of the human art that wants to represent it.

The baroque profusion of the branches is only apparently superfluous, since the message transmitted lies precisely in this abundance, and no leaf or figure or grape-cluster can be added or taken away. In other words, who the characters are and what they are called matters only up to a point: what counts is what is achieved through them.

The Tule Tree, a natural product of time, and the Tree of Jesse, a product of the human need to give some finality to time, are only apparently derived from a common scheme. Coming across them on the same day during my tourist journey, I feel that between them stretches the distance between chance and design, probability and determination, entropy and the sense of history.

A genealogical tree that wanted to render genuinely that process of procreation and death that is human survival should resemble not so much a tree of Jesse as a real tree, with its twisted and unharmonic ramifications, its stumps, its dry and green patches, the pruning that has come about by chance or through history, its waste of living matter. Actually, it ought to resemble the Tule Tree itself, where it is not clear what is root, trunk or branch.

But genealogical trees are always simplifications after the event, which privilege one particular line, usually a royal title or family name. In the tourist shops of certain French castles, they sell genealogical trees of the French kings, so that the tourists can orientate themselves in the complicated events which those places have witnessed. From the common stock of the Capets descend the lines of the Valois on one side and the Bourbons on the other, with the various Angoulême and Orléans branches as secondary ramifications, in a tree shape that is highly forced and asymmetrical.

An authentic genealogical tree should extend its own ramifications as much towards the present as towards the past, because at every wedding what should be represented is the joining together of two plants, and this would produce a highly intricate tangle expanding on all sides, and only coming to a halt in the irregular fringes where lines become extinct. It would be a bush whose ramifications at times expand, at times contract, because in a certain geographical area the same families all get mixed up again at each wedding. Would the form of the tree be renewed as one goes back to the roots of the human race, as to Adam and Eve in Christian iconology? For contemporary anthropology these roots are to be sought further and further afield, at a distance of millions of years, and they are scattered throughout the continents. (What seems to be approaching is the end, the severing of all the branches one by one or all together, the looming threat of a catastrophe that is demographic, nutritional, technological . . .)

The Forest and the Gods

In Palenque the soaring temples built on steps stand out from the background of the forest that rises above them with dense trees that are even higher than the temples: *ficus* trees with multiple trunks that look like roots, *aguacetes* with their shiny leaves, cascades of creepers, dangling plants and lianas. It seems that the forest is about to swallow up these colossal remains of the Mayan civilization; or rather, it swallowed them up many centuries ago, and they would be buried beneath a living and proliferating mountain of green were it not for the sharp blades wielded by men who, from the time those temples were discovered, have been fighting back on a daily basis against the assault of the vegetation to allow the stone constructions to emerge from the suffocating tangle of branches and shoots.

The bas-reliefs that the ancient Mayans sculpted in their stone-work represent, through figures of gods, stars and monsters, the various phases of the plant life cycle of maize. This at least is what is explained in books; what we can ascertain at first sight are sequences of signs in the shape of leaves, flowers or fruits, a vegetation of orna-mentation which flourishes around every vaguely anthropomorphic or zoomorphic outline, transforming it into an intricate tangle. Thus, whatever they mean, it was always vegetal forms that the Mayans fixed on their stonework: everything they constructed always came back to the flow of lymph in plants; a relationship that could almost be described as a mirror image has been established between the sculpted stone and the forest. The vegetal jumble becomes even more tangled in my head, which has been stunned by the sun and by the vertigo caused by climbing and descending those steep stairways, and amidst the ramifications of topics every now and again I seem to glimpse a decisive, over-arching reason, which disappears one minute later.

The bas-reliefs and the forest define and comment on each other in turn; the language of stone recounts and discusses the vital process that surrounds and determines it. But what sense does it make to say the word 'forest' when the actual forest is there, present and looming? If 'forest' is the word that is written in the sculpted figures of gods and monsters, then the temples in the forest are nothing but a giant tautology that nature rightly tries to cancel out as superfluous. So things rebel at the destiny of being signified by words, they reject that passive role that the system of signs would like to impose on them, and they recover the space that has been usurped; so they submerge the temples and bas-reliefs; once more they swallow up language, which had tried to assert its own autonomy and to establish its own foundations as though it were a second nature. The bas-reliefs telling stories of serpents, feathers and leaves disappear as they are invaded by nests of serpents and birds and tangles of liana. Language's dream of turning itself into a system and cosmos has been pointless: the last word belongs to silent nature.

This would be a neat conclusion, except that the same line of thought could also lead to an opposite inference. The forest can hammer away all it likes at the temples; but the stone does not allow itself to be corroded by the rotting of vegetal mucus; the figures where we can read the names of the gods refuse to be cancelled out by lichens and fungus. Ever since language has existed, nature has been unable to abolish it: it continues to operate despite everything, in its separate realm, which is not even touched by the convulsive attack of things. The names of the gods and the gods without names face up to each other in a war that cannot have either victors or vanquished.

But if I attribute an aggressive intention to the forest, if I see the roots and the lianas taking action, assaulting, outmanoeuvring the enemy, I am not doing anything other than projecting the mythology of the bas-reliefs on to what is vegetation and lymph. Language (every language) constructs a mythology, and this way it has of being mythological involves also what we thought existed before language. From the moment language made its entrance into the universe, the world has taken on language's way of being, and cannot manifest itself except by following its rules. From that moment on the roots and lianas became part of the discourse of the gods, from which every discourse stems. The actions made of nouns and verbs and consequences and analogies have involved the elements and prime substances. The temples that guard the origins of language atop the stone steps or at the bottom of subterranean crypts have imposed their dominion over the forest.

But today, are we sure that the gods still talk the language of the forest, from their ruined temples? Perhaps the gods that rule over discourse are no longer those that repeated the story, horrific but never leading to despair, of the succession of destruction and rebirth in an endless cycle. Other gods speak through us, conscious that what finishes never comes back.

Iran

The Mihrab

A frame sculpted in relief, surmounted by an architrave with a decoration perforated like a piece of lace; within the frame, a pierced decoration runs along the jambs with arabesques sculpted in low relief, and above all this, on a horizontal level, a line of fluent writing stands out as though suspended there. Everything is of the same light colour; the material is stucco. Below the frame emerges a tympanum with a pointed arch, framed by a grooved archivolt, upheld by thin columns, and thick with sculpted letters. In the margins every spare piece of surface is studded with ornaments, inlaid with lines and loops, and porous like a sponge. The columns and the positive ogive of the tympanum act as a frame for the negative ogive of a pointed arch, surmounted by a high architrave, which is also perforated, and the background to all this is hollowed out and sculpted in minute detail. At this point one would really need to use all the previous words again in order to describe similar details that are on a tiny scale and bunched and woven together in different configurations. And inside this arch that is at the very interior of all the arches, what does one see? Nothing: the bare wall.

I am trying to describe a fourteenth-century mihrab in the Friday Mosque at Isfahan. The mihrab is the niche inside mosques which indicates the direction of Mecca. Every time I visit a mosque, I stop in front of the mihrab, and never tire of looking at it. What attracts me is the idea of a door that does everything to put on display its function as a door but which opens on to nothing, the idea of a luxurious frame as though to enclose something extremely precious but inside which there is nothing.

In the Sheikh Lotf Allah Mosque the mihrab (from the

seventeenth century) is in a wall entirely covered with indigo and turquoise majolica, under an ogival span with at its centre a fake ogival window made of bright tiles that are criss-crossed by a geometric blossoming of spiral lines. The mihrab is a cavity – ogival again – which opens into the depth of the wall, which is resplendent with blue and gold majolica, and which is adorned over all its surface area with patterns of arches – hexagonal ones, this time. It has a vault composed of so many little cavities like a honeycomb, little cells without a floor which lie on top of each other in layers. It is as if the mihrab, by subdividing its own limited and composed space into a multiplicity of ever smaller mihrabs, was opening up the only way possible for it to reach the limitless.

All around, the white script flows across the blue tiles, binding the space with its calligrams, which are rhythmically punctuated by parallel bars, curves that loop like whips, speckles of oblique or dot-formed lines, launching the verses of the Koran on high as well as down below, to right and left, forwards and backwards, along every dimension that is visible and invisible.

After staying for a good while contemplating the mihrab, I feel I need to reach some conclusion. Which could be this: the idea of perfection which art pursues, the wisdom accumulated in writing, the dream of satisfying every desire that is expressed in the luxury of ornaments, all these point towards one single meaning, celebrate one foundational principle, entail one single final object. And this is an object which does not exist. Its sole quality is that of not being there. One cannot even give it a name.

Void, nothingness, absence, silence are all names that are heavy with meanings that are too cumbersome for something that refuses to be any of these things. It cannot be defined in words: the only symbol that can represent it is the mihrab. In fact, to be more precise, it is that something that is revealed not to be there at the end of the mihrab.

This was what I thought I had understood in that distant journey of mine to Isfahan: that the most important things in the world are the empty spaces. The honeycomb vaults of the cupolas of the Mosque of Shah Abbas; the dark cupola of the Friday Mosque which is supported on a succession of arches of decreasing size, calculated according to a sophisticated arithmetic in order to join the squared base to the circle supporting the canopy; the iwans, the great quadrangular doors with their arched vault: everything here confirms that the real substance of the world is provided by what is hollow.

The void has its own fantasies, its own games: the 'music room' in the Alì Qapù palace is covered along its walls and on its vault with an envelope of perforated, ochre-coloured chalk in which outlines of cruets or lutes are engraved in negative, like a collection of objects reduced to their own shadow or their idea of themselves without a body.

Certain forms of time are made for certain forms of space: the sunset hour in spring goes with the madrassa known as 'the Shah's Mother's Madrassa', an eighteenth-century enclosed garden, white with majolica and green with plants and ponds, above which there soar great raised rooms, empty, decorated by strips of tiles in which the agility of the writing comes to rest in the impassivity of the enamels. While visiting the madrassa, seeing the tranquil familiarity with which Isfahan's inhabitants feel this place and this hour, I think that I too would like to occupy the mezzanine of one of those spacious niches, like the man there who is sitting with his legs folded under him and reading, or the others who are chattering, or like that man who has stretched out and is sleeping, or like that other man who is eating bread in thin strips with salad. I envy the group listening to a mullah, as though they were Socrates' disciples, all crouching round one carpet, or the boys who have come out from school and are opening books and homework notepads on another carpet.

Perhaps a city that has been made following a happy arrangement of solid and empty space lends itself to being lived in with a cheerful spirit even in times of megalomaniac despotism: this was the thought that came to me as I walked in the animation of the evening, through the famous square in Isfahan, watching the mosques with their blue and copper cupolas, the houses all the same height, with their communicating terraces, and the wide vaults of Abbas the Great's palace and of the bazaar.

Some years have gone by. What I see now from Iran are very different images: with no empty spaces, it is full of crowds shouting and gesturing in unison, darkened by the blackness of the cloaks, which extends everywhere, full of a fanatical tension that knows no respite or peace. I saw nothing of all this when I was contemplating the mihrab.

The Flames within the Flames

The fire is preserved in the sacred chamber of the Zoroastrian temple, which is locked. Only the *mobet* has the key and can enter; during the ceremony the flame is visible through the iron grating.

The temple is a small, modern villa, surrounded by a modest garden, in Yazd, a city on the edge of the desert, in the centre of Iran. The *mobet* is a young Parsee Indian from Bombay (for more than a thousand years the Parsees in India have kept alive the most ancient religion of their ancestors who fled from Persia after the Islamic conquest); handsome, proud, with an attitude bordering on smugness; the white shirt he wears, the little white cap on his head, the white veil that covers his mouth to stop the sacred fire being contaminated by human breath all give him the look of a surgeon. He revives the fire with his little shovel; he adds some bits of sandalwood to the brazier. He recites the prayers to Ahura Mazda in a chanting voice, which begins in a whisper and slowly gets louder until it reaches top volume; then he stops, is silent, strikes a bell that resounds with deep vibrations. His voice alternates with the litanies of the women who are gathered in the temple, their heads covered by short, coloured mantillas, absorbed in the reading of their little books: prayers in a modern language, or at least one that is understandable nowadays, while the *mobet* prays in the Avestan language, in which are preserved the most archaic stratifications of the Indo-European language stock.

Is it to gather an echo of the mythical origins of words that I have come here, amongst the latest custodians of a discourse that has been handed down identical in letter and even accent for thousands of years? Or is it to see if something distinguishes from all the other fires the fire that apparently has been burning from the time of Cyrus, Darius, Artaxerxes, constantly rekindled from an uninterrupted succession of coals that have never been allowed to go out, a fire that has been guarded in secret during the 1,300 years of Islamic domination, and fed with seasoned and split sandalwood always according to the same rules, so as to produce a clear flame without a hint of smoke?

My journey to Iran is taking place in the last phase of the Shah's rule. This Shah persecutes many categories of people but not the minority that is faithful to the Mazdean religion (those whom we call Zoroastrians or Zarathustrians or, less accurately, 'fire-worshippers'). In opposition to the predominant Shiite clergy and from the time of the arrival on the throne of the present Shah's father, the Pahlevi dynasty has declared itself to be secular and tolerant of minority religions. Thus the capricious logic of political balances returned freedom of practice to the cult of Ahura Mazda, a cult which not only in its Indian exile but also in these remote regions of Persia had continued for centuries to be practised in secret, around fires that were always kept lit on the mountains and in houses.

With all the wariness of those who live amongst infidels, the Mazdeans continue to keep the fire locked away, visible only through a grating. But even when the altars flamed high on the monumental steps of Darius' Persepolis, the true chamber of fire was always a room without windows, aerated solely by air-holes and inaccessible to the sun's rays. There the flames were nourished with trunks of sandalwood seasoned to the point where every residue of earthly sap had disappeared, with the fire going out and being relit a thousand times from its own ashes. In this way the flames were purified of the dross of evil which pollutes all the elements and stars and plants and animals and above all man. The sacred fire shines in the dark: it must not mix its light with the light of day, which is exposed to all kinds of contamination. And perhaps even a human glance is enough to profane it, if it rests on the fire with indifference, as though it were a thing that was on the same level as all other things; like my glance, which is that of a man who vainly tries to recover a meaning for ancient symbols in a world which consumes everything it sees and hears. The true fire is the hidden fire: was it to learn this that I have come here?

Searching for the Zoroastrians of Yazd, yesterday afternoon we went back and forward across an endless, semi-deserted district, amidst blind walls made of earth and straw or of bricks of raw clay, terraces on the low, flat roofs from where a girl looks out, clusters of old women sitting around a thin threshold or underneath a niche in which a candle is burning. The women's religion is recognized by the shawl they cover their heads with: in this district the coloured ones outnumber the black ones. Through a door, a hallway, a series of communicating courtyards, we reached a low room where many candles were burning in front of some photographs of the dead: a kind of chapel, a space for a private cult; the fire, the famous fire, is announced only by these feeble little flames. The courteous passer-by whom we called on in the street and who has taken us this far gives us explanations that are lost on us because of the lack of a common language. He is even prepared to accompany us to the main temple, but only to show us that it is closed and that he can only point it out to us through the gate: an anonymous, modern, building. When we asked around, we learned that the next day they were expecting a foreign television crew, to film the celebration of a rite.

At the local office of the state television service, which is where we turned to, a functionary with five portraits of the Shah hanging on the wall or framed on his desk (the Shah on the throne, with his wife, with his children, in colour, in black and white) finds the contacts for us so we can be present at the filming.

Here I am, then, admitted to the temple, after I too put on a little white beret and took off my shoes (hair and the soles of shoes are the vehicles of contamination which one must guard against most), but everything I see still seems very distant to me. Distant from what? What have I come here to find amid the faithful followers of Ahura Mazda, the first god to reveal himself to the Indo-Europeans as the supreme transcendental principle? What can that mean for me, that bearded outline flanked by two huge wings which is repeated everywhere, from Darius' bas-reliefs at Persepolis to the modest modern furniture in this little room? He is a schematic human figure seen in profile, with a long, curly beard and hair similar to his beard and on top a cylindrical hat: in his hand he holds a circle and he in turn is surrounded by another, bigger circle, from which there open out two huge wings, maybe eagle's wings, and some forewings or antennae which are perhaps lightning-bolts; only the figure's bust is visible, down to his waist, framed by the winged circle like an aviator installed in the cockpit of a primeval flying machine. It would be natural to believe that this is Ahura Mazda in person, but I certainly will not fall into such a vulgar error, because I know there can be no images of an invisible, omnipresent god (just as Ahura Mazda is also just a way of speaking, not a name). At most he must be a divine emanation, which descends from heaven on to the heads of Emperors, or a heavenly archetype of their majesty, and which we instead can understand as hovering above us, a benediction to invoke or a model to imitate.

In short, Ahura Mazda remains distant, even in this temple with its neon lights, and the metal chairs painted white, and the white-robed priest who is very happy to officiate in front of the television cameras. There are few decorations hanging on the wall: a painting showing Zarathustra in the style of those popular Oriental oleographs, a mirror, a calendar in which the emblem of the bearded man with wings stands out against the Iranian tricolour.

The only image possible of Ahura Mazda is fire. Shapeless, limitless, it heats and devours and spreads, with the agility of its dazzling tongues, which change colour every second: the fire that languishes in its slow death in the brazier, which hides itself under the grey ash, and suddenly flares up again, raises its pointed wings, recovers its impetus, soars upwards in a violent burst of flames. All I have left to do is to stare at the glare of the flame rising up from the hidden brazier, and to look at the men and women praying to the fire and to try to imagine how they see it. With attraction and fear, as I see it? Certainly: as a friendly force, a necessary condition of our existence, but the attraction that the sight of the flames exercises is more instantaneous than any reasoning about it, it is instinctive like the terror that the sight of fire instils in us as an enemy force, a force for destruction and death. And even further beyond that they see in the fire an element that is incompatible with everything that is obliged to be subject to the business of life and death, an absolute way of being, so much so that they associate it with the notion of ideal purity. Perhaps because man may think he can master it but cannot touch it? Because inside it no living being can survive? Is what is untouchable by man pure? Is what excludes life from itself pure? Is what lives stripping itself of every body and wrapping or support pure? And if purity is in the fire, how can one purify the fire? By burning it? Is the flame that the Mazdeans are reciting their prayers to a fire that has been set on fire? Is it a flame that has been set on fire?

Over and over again the stars continue to burn their fuel through century after century. The firmament is made of braziers that light up and go out, incandescent supernovae, red giants that slowly die out, burnt-out relics of white dwarves. The earth too is a ball of fire that is expanding the crust of the continents and the ocean sea-beds. The universe is one big fire. What will happen when all the sandalwood of atoms has disappeared in the stars' crucibles? When the ashes of ashes are consumed in one blaze of evanescent heat? When the pyres of the galaxies are reduced to opaque vortices of soot? How can one conceive of a fire that keeps itself lit from the beginning of time and that never goes out?

The world I inhabit is governed by science, and this science has a tragic core: the irreversible process that will lead the universe to decompose in a cloud of heat. Of the liveable and visible worlds there will remain only a dust-cloud of particles which will no longer find a shape, where nothing will be distinguishable from anything else, the near and the distant, the before and the afterwards. Here amidst the faithful followers of Ahura Mazda, in the fire which has been guarded in the dark and which the *mobet* revives and nurses to the sound of his chanting voice, I am shown the substance of the universe which only manifests itself in the combustion which ceaselessly devours it, the form of space expanding and contracting, the rumble and crackle of time. Time is like the fire: at times it flares up in impetuous bursts of heat, at times it smoulders buried in the slow carbonization of epochs, at times it creeps and spreads out in unexpected, lightning-quick zig-zags, but it always points towards its only end: to consume everything and to be consumed. When the last fire goes out, time too will be finished; is that why the Zoroastrians perpetuate their fires? The thing I seem to be on the point of understanding is this: it makes no sense complaining that the arrow of time rushes towards the void, because for all that exists in the universe and that we would want to save, the fact of being there means just this burning and nothing else: there is no other way of being except that of the flame.

Who knows whether I could find in Avestan a formula to express these thoughts? For the moment, going back to my Western memory, I find the remark of a poet enough. To whoever asked him this question: 'If your house was being destroyed by fire, what thing would you rush to save?' Jean Cocteau replied: 'The fire.'

The Sculptures and the Nomads

At Persepolis, I find myself going up the monumental staircase along with two lines of people forming two columns: a row of tourists all in groups and a line of dignitaries with curly beards and curly hair, with cylindrical coiffures interwoven with feathers, massive half-moon necklaces around their neck, sandals on their feet underneath their pleated togas, and sometimes a flower in their hand. The first row is made of flesh and blood and sweat, the second of sculpted stone. Allowing the first line to go ahead under the burning sun, I empathize with the uninterrupted gait of those dignified figures on the grey surface of the stone slabs, with that solemn procession which advances wherever one rests one's gaze on all the stairs of the city, along the base of all the façades, as it flows towards the doors flanked by winged lions and then the hall of a hundred columns. The stone population is of the same size as that of flesh and blood, but is distinguished by its composure and a certain uniform rigidity in lineaments and dress, as though it were the same figure in profile that constantly passed by. Every so often a face looks back at the person behind, a hand is placed on the chest or on the shoulder as if in a gesture of friendship, introducing a note of animation into the ceremonial formality, an animation that is all the warmer the more stereotyped the hieratic nature of the rest of the procession seems.

The palace of the Achaemenid kings at Persepolis is like a container which reproduces on its walls what went on inside it 2,500 years ago. Its architecture was made for displaying a sumptuous procession which could not but reproduce the kind of ceremonies that had always gone on there, in every grouping and in every gesture, in the arrangement and succession of every embassy and every group, in the display of costumes, wealth and weapons: the imperial guard with lances, bows and quivers, the gift-bearers from various nations with precious vases and little bags of gold-dust.

In the bas-relief on the main door, the nations support the imperial throne, but this throne is so light that they can hold it up with their fingertips. Or to be more precise: above the great throne that the ambassadors of the nations raise up, barely touching it underneath its cross-beams, there is a smaller throne, on which sits a little emperor flanked by a slave with a fly-swat, and above him is a canopy, and above that again hovers the emblem of Azura Mazda or of his benediction. Now one begins to understand where all those processions converging on the doors, vestibules and access corridors are going: the more one approaches the centre of power the more one moves from the enormous to the tiny, the reduced; to abstraction, the void. Perhaps this palace is the utopia of the perfect empire: a great empty box ready to receive the shadows of the world, a procession of figures in profile, flat figures, with no depth, around an empty, weightless throne.

Other crowd scenes are on display a few kilometres from here, on a sheer rock in the Naqsh-e Rustam gorge, but these are battle scenes with horses trampling enemies who have been unseated, the threatening armour of warriors lined up on the battlefield, prisoners made slaves and weighed down by chains, triumphs and divisions of spoils. It was the Sassanid kings who had these rocks sculpted to celebrate their own achievements, more than 500 years after the destruction of Persepolis, immediately beneath the tombs of their ancient Achaemenid ancestors: Darius, Artaxerxes, Darius II, buried behind four austere blocks like palace façades sculpted on a high ledge above the cliff. The composed, rapt majesty of Persepolis has disappeared: here what dominates is pride, bellicosity, the affirmation of one's superiority over the enemy, the ostentation of opulence. It is a humanity on horseback recording its way of life for posterity. It is an epic of attacks at full gallop, the apotheosis of equestrian supremacy, with the din of trumpets, clouds of dust and the echo of hooves on the earth: all this is recorded in the shapes that emerge from the rocks. An elegant Shapur I, all frills and necklaces, lifts his arm and his sword from on top of a powerful horse at whose feet the defeated Roman Emperor Valerian is kneeling, his hands outstretched and trembling, his eyes filled with terror. Even before that Azura Mazda in person offers to Ardashir I the diadem of investiture from which dangle long thin ribbons. For the first time the god is visible: and he is a knight the same size as the Sassanid king, dressed with equal pomp, mounted on a horse that is just as powerful.

On the way back, my route intersects with that of a tribe of nomads on the move. Barefoot women, with garish-coloured clothes, are chasing forward a row of little donkeys, beating them with sticks and yelling. On some donkeys' backs are balanced a hen, a dog, and a lamb astride the donkey; others have saddlebags from which lambs and new-born babies stick out. The last little donkey trudges along: on its back sits an old witch, roaring, riding side-saddle, with a stick in her hand; all the kinetic energy that pushes the caravan forward seems to emanate from this old woman. This is followed by a herd of goats, then a herd of camels; a little white camel trots in between its mother's legs. The procession heads towards an encampment of black tents. This is the season when the tribes of these Turkish-speaking nomadic populations cross the steppes of the land of Fars; after wintering on the shores of the Persian Gulf they go back north every year towards the Caspian Sea. Unlike the women, the men are dressed like city-dwellers; they wait at the threshold of their tents, greet foreigners with a *Salam!* and invite them in to drink tea. At the arrival of these strangers some of the women hide their faces and laugh in the black and white of their eyes; one of them pours water from a goatskin water-bag; another starts to knead the dough. On the ground are the famous carpets woven on their looms. For centuries the nomads have criss-crossed these arid terrains between the Persian Gulf and the Caspian Sea without leaving any trace of themselves behind apart from their footprints in the dust.

In one single day I have done nothing but meet human crowds on the march crossing my path: rows of people fixed for ever in the rock and other rows of people who are on the move in perpetual transit. Both inhabit different spaces from our own: one lot merges with the compact mineral world, the others barely graze places, ignorant of the names of geography and history, following itineraries that are not marked on any map, like the migrations of birds. If I had to choose between the two ways of being, I would have to weigh up their pros and cons for a long time: either living only in order to leave behind an indelible sign, transforming oneself into one's own figure engraved on the page of stone, or living by identifying with the cycle of seasons, the growth of the grasses and bushes, with the rhythm of the years that cannot stop because it follows the revolutions of the sun and the stars. In each of these cases what one is trying to escape is death. In each of these cases it is immutability that one is aiming for. For one group death can be accepted as long as what is saved from life is the moment that will last for ever in the uniform time of stone; for the others death disappears in cyclical time and in the eternal repetition of the signs of the zodiac. In each case something holds me back: I cannot find the gap where I could insert myself and join the crowd. Just one thought makes me feel at ease: the carpets. It is in the weave of their carpets that the nomads deposit their wisdom: these variegated, light objects are spread on the bare ground wherever they stop to spend the night, and are rolled up again in the morning so they can carry them away with them along with all their other belongings on the humps of camels.